Biochemical & Medicinal Chemistry Series

Series Editor
JOHN MANN
Department of Chemistry, University of Reading

Titles in this series

Neuropharmacology
T. W. STONE

An Introduction to Biotransformations in Organic Chemistry
JAMES R. HANSON

Bacteria and Antibacterial Agents
JOHN MANN AND M. JAMES C. CRABBE

Biochemical & Medicinal Chemistry Series

An Introduction to Biotransformations in Organic Chemistry

JAMES R. HANSON

School of Chemistry and Molecular Sciences
University of Sussex

W.H. FREEMAN
SPEKTRUM
OXFORD · NEW YORK · HEIDELBERG

W. H. Freeman and Company Limited
20 Beaumont Street, Oxford OX1 2NQ
41 Madison Avenue, New York, NY 10010

British Library Cataloguing in Publication Data
A catalogue record for this book is available from the British Library.

Library of Congress Cataloging-in-Publication Data
Hanson, James Ralph.
An introduction to biotransformations in organic chemistry / James R. Hanson
p. cm. — (Biochemical and medicinal chemistry series)
Includes bibliographical references and index.
ISBN 0–7167–4541–0
1. Biotransformation (Metabolism) 2. Organic compounds–Synthesis.
I. Title. II. Series.
QP517.B58H36 1995 574.19'24—dc20 95–7352 CIP

Set by Keyword Publishing Services
Printed by Bell & Bain Limited

Contents

Preface

The study of the transformation of exogenous compounds by living systems has antecedents that date back to the work of Pasteur in the last century. Since then biotransformations have played a crucial role in several classic syntheses, but it is only in recent years that they have begun to be widely exploited in synthesis. The stimulus for these applications has been the enantioselectivity of many biotransformations. Furthermore, biotransformations can be carried out under simple and mild reaction conditions and with a regioselectivity that sometimes usefully complements chemical reactivity. The synthetic chemist must be able to assess when it is appropriate to explore the potential of a biotransformation.

In this book, the use of both isolated enzyme systems and whole organisms in biotransformations is considered. The objective is to provide an introduction to the subject showing those situations in which biotransformations may present viable alternatives to purely chemical synthetic routes or be used in a complementary manner to chemical reagents in a multi-step organic synthesis. This is not a textbook of enzymology nor does it aspire to be a research monograph. A number of research monographs and compendia of experimental methods, to which the reader may turn, are listed at the end of the book.

I wish to thank Professor Sir John Cornforth A.C., F.R.S. and Professor John Mann for reading the manuscript and for their helpful comments.

1 Introduction

The use of biological methods to bring about chemical reactions forms a bridge between chemistry and biochemistry. For the purposes of this book, biotransformations are defined as the use of biological systems to bring about chemical changes on compounds that are not their natural substrates. This distinguishes biotransformation from biosynthesis, which is concerned with the natural synthetic ability of biological systems in their normal habitat.

It is possible to distinguish between various approaches to biotransformation. There are xenobiotic transformations that involve the biotransformation of substrates that are completely alien to the particular system. In contrast, there are biosynthetically directed biotransformations in which the substrate bears a formal relationship to a natural biosynthetic intermediate. Whilst both may be used for synthetic purposes, the latter can also shed some light on structural and mechanistic features of biosynthesis.

Isolated enzyme systems or intact whole organisms may be used for biotransformations. Each approach has its advantages and disadvantages. Many isolated enzyme systems are now commercially available or are relatively easy to isolate, at least in a crude form. They can be stable and easy to use, often giving clean, single products. For many hydrolytic reactions, no co-factor is needed. However, for redox reactions in which a co-factor is used, the need to regenerate this can add a complication. Whole organisms do not have this disadvantage. They do tend to give more than one product, which may or may not be an advantage. They are often cheaper to use than isolated enzyme systems.

Biotransformations have a number of advantages when viewed alongside the corresponding chemical methods. Many biotransformations are not only regio- and stereospecific but are also enantiospecific allowing the production of chiral products from racemic mixtures. The conditions for biotransformations are mild and in the majority of cases do not require the protection of other functional groups. Furthermore, the features governing their regiospecificity differ from those controlling chemical specificity and indeed it is possible to obtain biotransformations at centres that are chemically unreactive (e.g. (**1**) transformed to (**2**)). The dominant feature in a biotransformation is the topological relationship between the substrate and the active site of the enzyme. From a commercial point of view some biotransformations can be cheaper and more direct than their chemical analogues whilst the transformations proceed under conditions that are normally regarded as 'environmentally friendly'. However, many of the ground rules for applying biotransformations are not yet well-understood or well-defined and there are many chemical reactions for which there is no equivalent biotransformation. Thus, the chemist has to learn to use biotransformations alongside the conventional chemical reagents.

History

One of the first applications of biotransformations was by Pasteur, who reported in 1858 the use of the fungus *Penicillium glaucum* to obtain L-ammonium tartrate from DL-ammonium tartrate by selective destruction of the D-enantiomer – the first example of a kinetic resolution. Pasteur also turned his attention to another biotransformation, that of ethanol to acetic acid (vinegar) by *Acetobacter aceti*, an organism then described as *Mycoderma aceti* or *Bacterium xylinum*, publishing studies on this in 1864. These studies were extended by Brown who reported in 1886, the oxidation of propanol to propionic acid by an organism described as *Bacterium aceti*, which also carried out the oxidation of ethanol. This paper also reports a continuation of earlier studies by Berthelot on the biotransformation of mannitol by the same organism to laevulose (fructose) and of dextrose (glucose) (**3**) to gluconic acid (**4**). However, many of these early studies were hampered by the lack of pure cultures. The reducing action of yeast, *Saccharomyces cerevisiae*, was reported by Dumas in 1874, who described the conversion of sulfur to hydrogen sulfide. This was followed by the demonstration in 1898 by Windisch that furfural (**5**) could be reduced to furfuryl alcohol (**6**). The term 'phytochemical reduction' was coined for these transformations.

The early part of this century saw the development of many biotransformations based on the use of the yeast, *Saccharomyces cerevisiae*, particularly under the direction of Liebig and Neuberg. Some of these biotransformations arose from studies directed at the microbiological production of acetone, glycerol and butanol during the First World War. A particularly useful biotransformation was developed by Neuberg in 1921. This was a chiral acetoin condensation mediated by yeast. The addition of benzaldehyde (**7**) to the fermentation gave the ketol (**8**) by condensation with endogenous acetaldehyde. This led in 1934 to a commercial synthesis of the alkaloid, ephedrine (**9**).

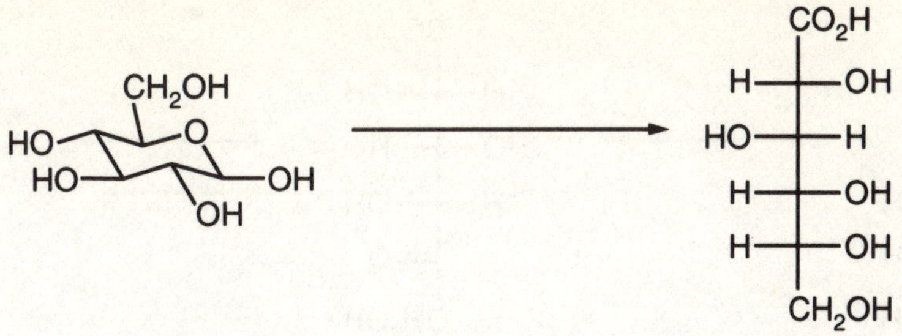

3

4

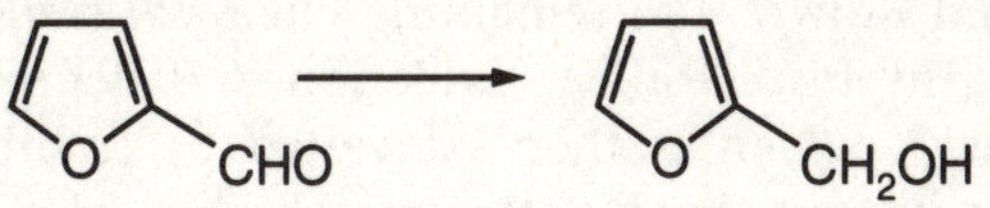

5

6

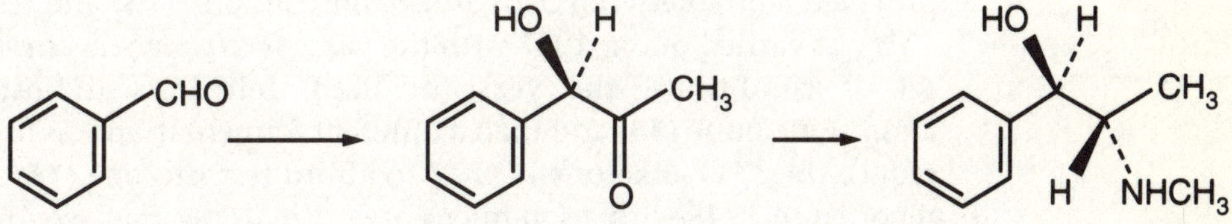

7

8

9

The reductive alkylation of the oxyanions of arsenic, selenium and tellurium by various Penicillium species was described in 1933 in the context of the long-standing problem of the formation of volatile toxic organo-arsenic compounds from arsenic-containing pigments.

During this period a number of biosynthetic and general metabolic pathways were indicated by examining the biotransformation of putative intermediates and compounds related to them. The metabolic interrelationship between some of the simpler carboxylic acids and aminoacids and the glycolytic pathways were established by these methods before the advent of isotopic labelling. The biodegradation of aromatic compounds, including the formation of muconic acids, was also established. Some enzyme systems were separated. The foundations of some of the biotransformations that are proving valuable today were laid in the work carried out during the 1920s to 1940s.

A useful industrial application developed during the 1930s was the synthesis of vitamin C (**12**) based on the work of Reichstein published in 1934. This utilized the bacterial oxidation of D-sorbitol (**10**) to L-sorbose (**11**) by *Acetobacter suboxydans* (*Acetobacter xylinum*) and was based on the earlier observations of

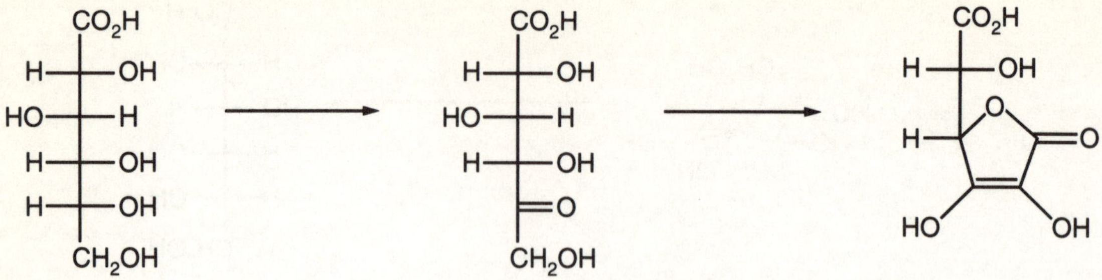

Pasteur and Brown. The remaining chemical steps involve protection of L-sorbose as its bisacetonide, oxidation and deprotection to form 2-keto-L-gulonic acid, which is then converted to ascorbic acid (vitamin C). An alternative process involves the microbiological oxidation of D-glucose with *Acetobacter fragrum* to 2, 5-diketo-D-gulonic acid and then the selective reduction of this by a Corynebacterium species to 2-keto-L-gulonic acid.

During the structural work on the steroid hormones, a biotransformation provided a link between dehydroisoandrosterone (**13**) and testosterone (**15**). This was carried out in 1937 with the yeast *Saccharomyces cerevisiae.* Under one set of conditions, the yeast oxidized dehydroisoandrosterone (**13**) to androstenedione (**14**) and then a different fermentation was used to selectively reduce the 3, 17-diketone at C-17 to afford testosterone (**15**). Further studies reported in 1948 with a soil micro-organism, (*Proactinomyces roseus*) described the degradation and oxidation of sterols including the oxidation of cholesterol (**16**) at C-7. This was exploited in 1952 when a fungus, *Rhizopus arrhizus*, was found that was capable of hydroxylating progesterone (**1**) at the biologically important C-11 position to afford (**2**). The development of the efficient microbiological hydroxylation of steroids to provide intermediates for the synthesis of the corticosteroids, was a significant stimulus to the general study of the applications of biotransformation. The progesterone for this work was obtained from the plant steroid, diosgenin, the side chain of which could be degraded by chemical routes. However, problems of supply led to the investigation of other starting materials. Plant oils, such as soya bean oil, contain high proportions of sitosterol (**17**) and stigmasterol (**18**). Bacterial methods have now been developed for the degradation of these to androstenedione (**14**) from which the various steroid hormones can be prepared.

An important application of amidases involves the cleavage of the natural penicillins (e.g. (**19**)) produced by *Penicillium chrysogenum* to 6-aminopenicillanic acid (6-APA; (**20**)). This β-lactam is the parent substance from which the semi-synthetic penicillins, such as ampicillin and amoxocillin, are made. The amidase, obtained initially from *Bacillus cereus*, has been immobilized on a resin and the process has been turned into a high yielding and effective means of producing a key intermediate for antibiotic synthesis.

HO 13 14 OH 15

22 23

HO 7 16 HO 17

18 double bond C-22: C-23

A useful hydration, which has been scaled up to operate at the tonnage level, is the hydrolysis of acrylonitrile (**21**) to acrylamide (**22**) by the bacterium, *Rhodococcus rhodochrous*. The conditions are considerably milder than those required for the chemical hydrolysis. This process produces material for use in the polymer industry.

19 **20**

$$CH_2{=}CH.C{\equiv}N \longrightarrow CH_2{=}CH.\overset{O}{\overset{\|}{C}}.NH_2$$

21 **22**

Many drugs produced by total chemical synthesis are racemates. However, their beneficial effect often resides in one enantiomer. It has been realized, particularly since the thalidomide tragedy of the early 1960s, that the other enantiomer, far from being a harmless diluent, may actually be the source of adverse side-effects and toxicity. Hence, it is important in the pharmaceutical industry to produce single enantiomers. This has given an immense stimulus to the use of biotransformations in chiral synthesis. These developments are particularly seen in the application of enzymatic hydrolyses to produce chiral alcohols and acids and in the formation of *cis*-cyclohexadiene-1,2-dihydrodiols from aromatic compounds by oxidation with *Pseudomonas putida*.

The synthesis of chiral molecules by enzymatic methods has led in recent years to the widespread use of the hydrolytic properties of esterases and lipases. Consequently, predictive models of, for example, pig liver esterase, have been developed to delineate the scope of these transformations. Useful models that define the boundary conditions within which the structure of a substrate must fall, have also been developed for a number of redox enzymes such as horse liver alcohol dehydrogenase. The last 15 years have also seen a far better understanding of enzyme reaction mechanisms and this has had a valuable impact on the application of particular enzymes in synthesis.

In the past, the application of specific enzymes has been limited by their availability. Genetic engineering and the ability to transfer genetic information from one organism to another involving recombinant DNA techniques have changed this. Enzymes such as some of the aldolases involved in C–C bond formation and even plant enzymes responsible for alkaloid biosynthesis have been expressed in the bacterium *Escherichia coli*. The accessibility of a number of enzymes, as a result of genetic engineering, has removed a major obstacle to their use. There are reports of these cloned enzymes being purified and combined with others in single flask multi-enzyme syntheses.

An interesting development with an impact on biotransformation involves the production of catalytic antibodies. An antibody possesses the property of molecular recognition, by which it binds to specific molecular structures. Antibodies may be stimulated by the recognition of particular structural features in a macromolecule, the haptenic groups on a hapten. Consequently it is possible to raise antibodies that recognize stable fragments of a molecule, which possess a structural similarity to intermediates in chemical reactions. Hence an antibody can be developed with a binding site that can accommodate the transition state of a reaction and thus, by bringing the reactants together in a favourable way, these antibodies or abzymes possess some catalytic activity. For example, a tetrahedral phosphonate (**23**) mimics the tetrahedral intermediate (**24**) in ester hydrolysis. The catalytic antibody elicited by exposure to (**25**) will catalyse the hydrolysis of the ester (**26**).

A number of developments in biotechnology have had an impact on biotransformations. A problem with many enzyme-catalysed reactions is that the isolation of the reaction product can involve separation methods that may lead to the denaturing of the enzyme and restriction of its subsequent use. However, by attaching the enzyme to beads of resin, which can be filtered after use and

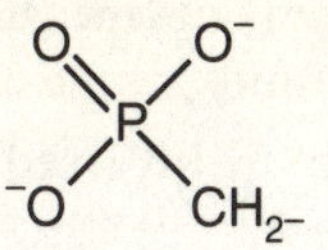

23

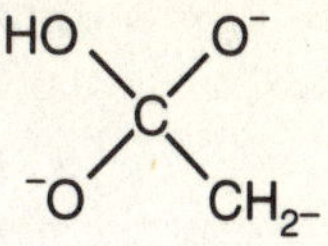

24

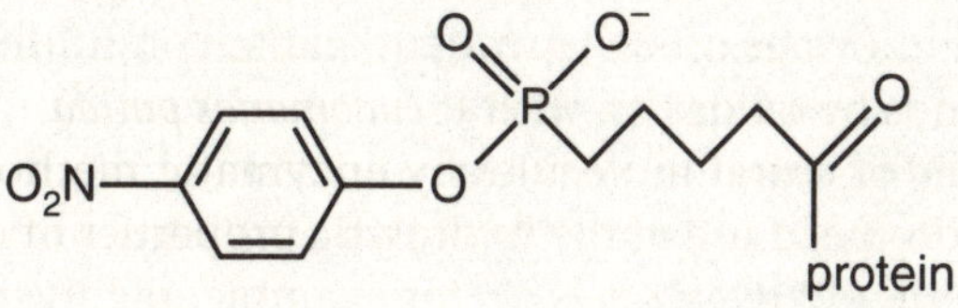

protein

25

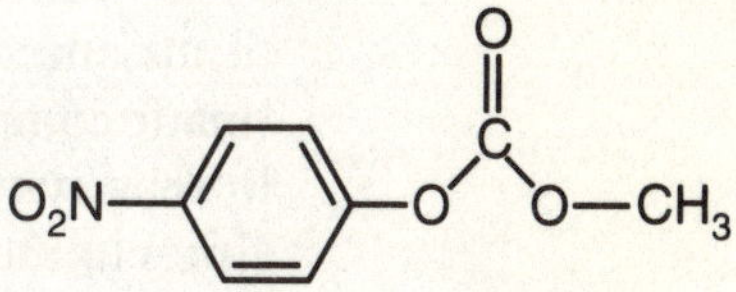

26

easily separated, the problem can be overcome. It is also possible to immobilize some bacteria on a solid support and so facilitate their separation. A further problem lies in the low solubility of many organic substrates in water. However, in 1985 it was shown that some enzymes, particularly lipases, could be used in organic solvents. This, and the use of biphasic systems, is widening the scope of many biotransformations.

Biotransformations are normally associated with isolated enzymes or with micro-organisms. However, many compounds of interest are produced by higher plants. Plant tissue culture has now reached the stage in which it is being used not only for the study of biosynthesis and the production of natural products but also examined for its potential in the biotransformation of unnatural substrates.

Classification of enzyme systems and units

There are many enzyme systems available and quite a number can be obtained from commercial suppliers. They have been classified by the Commission on Enzymes of the International Union of Biochemistry. The recommendations of the Nomenclature Committee of the I.U.B. on the naming and classification of enzymes are published from time to time. Enzymes are placed in the following groups.

1. ***Oxidoreductases:*** These enzymes mediate oxidation and reduction, including the insertion of oxygen into C–H and C–C bonds and the addition of oxygen to alkenes. This group also includes enzymes that are responsible for the addition or removal of hydrogen.
2. ***Transferases:*** These enzymes are involved in the transfer of one group, such as an acyl or a sugar unit from one substrate to another.
3. ***Hydrolases:*** This group includes the enzymes that mediate the hydrolysis or formation of amides, epoxides, esters and nitriles.

4. *Lyases:* These are a group of enzymes that fragment larger molecules with the elimination of smaller units.
5. *Isomerases:* These enzymes are involved in epimerization, racemization and other isomerization reactions.
6. *Ligases:* This group includes the enzymes responsible for the formation of C–C, C–O, C–S and C–N bonds.

In considering this classification, it is worth remembering that the majority of enzyme systems can be used, dependent on the conditions, to catalyse a reaction in both the forward or the reverse sense. At present most biotransformations using isolated enzyme systems employ hydrolases or oxidoreductases although there has been some use of enzymes from the other classes such as the transferases, lyases and ligases.

Within this classification, enzymes are divided into sub-groups indicating the substrate or reaction type. These sub-groups are further divided in terms of any additional co-substrate and finally the enzyme is given a number within this last group. Hence an enzyme may be classified with a four-component number: E.C. W, X, Y, Z, where W refers to enzyme class, X and Y to the sub-groups and Z refers to the individual enzyme.

The standard system for describing the catalytic activity of a preparation is the International Unit (I.U.). An International Unit of an enzyme will catalyse the transformation of 1 μmol of its substrate per minute. Other units such as nmol/min or the katal (1 kat = 1 mol s^{-1} of substrate transformed) are also sometimes used. However, it should be noted that the unit of catalytic activity is specified in terms of a particular substrate and set of conditions, which may not bear much relationship to the biotransformation for which the enzyme system is to be used. Another parameter that may be quoted is the catalytic constant. This is defined as the moles of product that are formed per minute per mole for pure enzyme. The turnover number is equal to the catalytic constant divided by the number of active centres in the enzyme. It will be identical to the catalytic constant if the enzyme is monomeric and each enzyme molecule interacts with one substrate molecule at a time. For further details the reader should consult one of the standard textbooks of enzymology.

Biotransformation experiments

The basic experimental methods that are used for biotransformations are no more difficult to carry out than those that are used, for example, to manipulate air-sensitive organometallic compounds. The experiments require careful preparation and a high standard of cleanliness to avoid contamination. In this context it is worth remembering that the spores of contaminating micro-organisms can be present in the air of the laboratory, on surfaces and on the skin of the chemist. Adventitious bacterial, yeast and some fungal infections can grow quite rapidly on a nutrient medium. Consequently, it is sensible to obtain a sterile cabinet for carrying out inoculations and to dedicate a clean laboratory for the purposes of carrying out biotransformation experiments.

The apparatus required for an enzymatic biotransformation is simple, involving conical flasks and a constant temperature bath. A typical procedure might involve dissolving or suspending the reactant in a buffer and stirring this with the enzyme preparation and, if necessary, any co-factors. The buffer is required to maintain the optimum pH for the particular enzyme. Enzyme reactions are particularly pH sensitive and if the reaction involves the release of an acid, it may be necessary to maintain the pH near neutrality by the addition of alkali. The concentration of the substrate may also be important. Some enzyme reactions proceed satisfactorily at 0.1 M but fail with a more concentrated solution. Other enzyme reactions may be inhibited by the product. It is also important to control the temperature. Some reactions may require a temperature of up to 35°C, particularly if the enzyme was of bacterial origin. The reaction, which may last for 24–48 h, should be monitored, typically by t.l.c., and on completion the reaction is worked up in a conventional chemical way. Enzyme reactions can be quite variable and if a chiral product is being generated, it is important to check the enantiomeric excess of each batch. In an enzymatic resolution it is sometimes useful to relate the effectiveness of the transformation to the enantiomeric excess observed in both the product and the residual starting material. A mathematical treatment devised by Sih exists for the expression of this in numerical terms.

Transformations using whole organisms utilize a different methodology and can require some simple microbiological skills. The organisms that are required may be obtained from the various culture collections. Although many organisms have been used for biotransformations, in practice a small group are often examined in the first instance for a particular biotransformation. The strain of an organism that has been used for a particular biotransformation should also be recorded. The organism may be maintained as a master culture on an agar slope. A number of sub-masters can be prepared from this. The composition of the medium on which the organism is to be grown will depend on whether it is a bacterium, yeast or a fungus. A typical aqueous medium would include a sugar, a nitrogen source (e.g. ammonium nitrate), phosphate and various mineral salts. This medium would be sterilized in appropriate vessels, e.g. conical flasks, in an autoclave and then a seed flask would be inoculated with the organism and grown for a few days. The bulk culture would be inoculated from this and the organism may then be grown for one to two days to establish itself prior to the addition of the reactant. The latter, typically one gram for five litres of medium, dissolved in a solvent such as ethanol, dimethylformamide or dimethylsulfoxide, would be evenly distributed under sterile conditions between the flasks. The incubation is then continued in shake or stirred culture for a period of two to ten days depending on the transformation. After the fermentation, the fungal mycelium is filtered from the broth and the metabolites are recovered both by extracting the broth and rinsing the mycelium with a suitable organic solvent such as ethyl acetate. Conventional chemical techniques are then used to separate, purify and identify the products.

In experiments that involve biosynthetically directed biotransformations, the separation of the metabolic products of the reactant may be facilitated by incorporating an inhibitor of an early stage of the biosynthesis into the medium. This can have the effect of preventing the formation of the endogenous natural products whilst allowing the transformation of the reactant along the later stages of the pathway. The selection of strains for xenobiotic transformations and the induction of the relevant enzyme systems can be facilitated by growing slopes or seeds of the organism on a limited medium containing the reactant. Interestingly, some useful strains of organisms have been obtained from soil contaminated by industrial waste related to the substrate under investigation.

2 Hydrolytic reactions

Esterases and lipases

Hydrolyses of amides and esters by enzymatic methods are now routine biotransformations. A range of proteases, lipases and esterases that will accept a variety of substrates are commercially available in varying degrees of purity and homogeneity. Furthermore, the general mechanisms by which some of these operate are quite well-known and their scope has been investigated to the extent that some predictive models are available. Nevertheless, caution must be exercised in the extent to which the conclusions obtained with one system may be extrapolated to another. These hydrolytic enzymes do not require additional co-factors and hence there are not the problems associated with co-factor recycling.

A major application of the esterases and lipases lies in the preparation of chiral molecules for synthesis. This presents a dichotomy, since the enzyme must accept a broad range of substrates whilst showing a high enantioselectivity for each one. For the most part these hydrolyses are kinetic resolutions rather than complete resolutions and some experimentation is necessary to establish the optimum conditions to obtain the highest enantiomeric excess. The chiral discrimination may occur in two senses. Firstly, it may involve merely the distinction between two enantiomers in a racemate. However, an increasingly valuable use involves a distinction between two paired ester groups (i.e. enantiotopic groups) attached to a prochiral center in a non-chiral molecule. Here chirality is generated and yields can by much higher than the 50% limit for a racemate. These applications are illustrated by Schemes 1 to 4.

Some of the commonly used enzyme systems are given in Table 2.1. The mechanism of action of the enzyme systems parallel those found in the chemical hydrolysis of esters and involve the addition of a nucleophile to the carbonyl group with the formation of a tetrahedral intermediate. This subsequently collapses with the liberation of the free alcohol. The nucleophile may be the hydroxyl group of serine as in the serine proteases, the carboxylate of aspartic acid as in pepsin or the thiol of cysteine as in papain.

Scheme 1

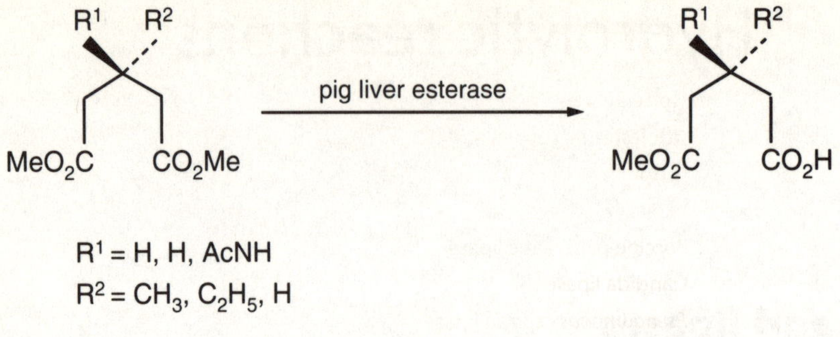
R1
R2
MeO2C
CO2Me
pig liver esterase
R1
R2
MeO2C
CO2H
R1 = H, H, AcNH
R2 = CH3, C2H5, H

Scheme 2

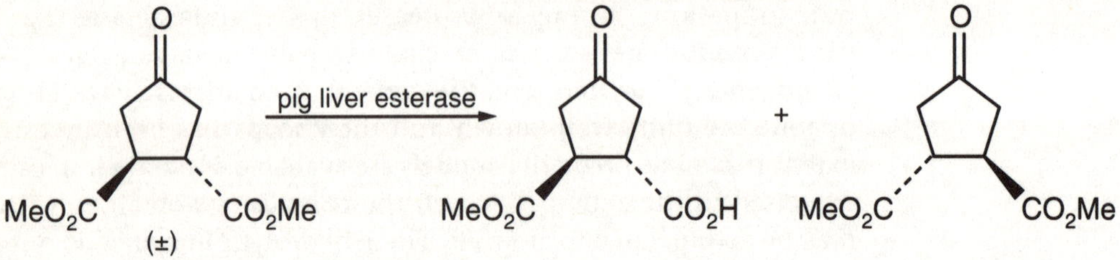
O
MeO2C
CO2Me
(±)
pig liver esterase
O
MeO2C
CO2H
+
O
MeO2C
CO2Me

Scheme 3

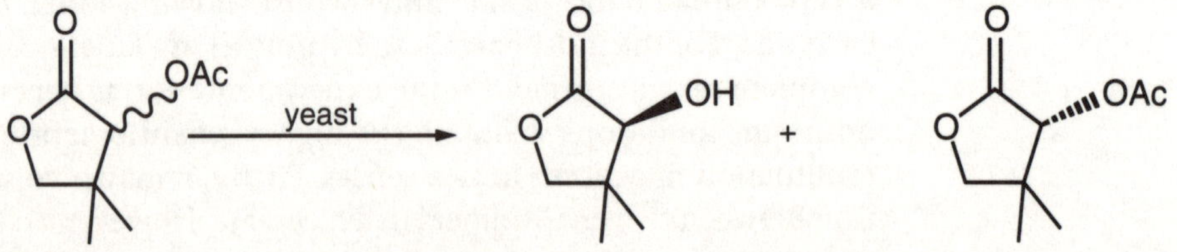
O
OAc
O
yeast
O
OH
O
+
O
OAc
O

Scheme 4

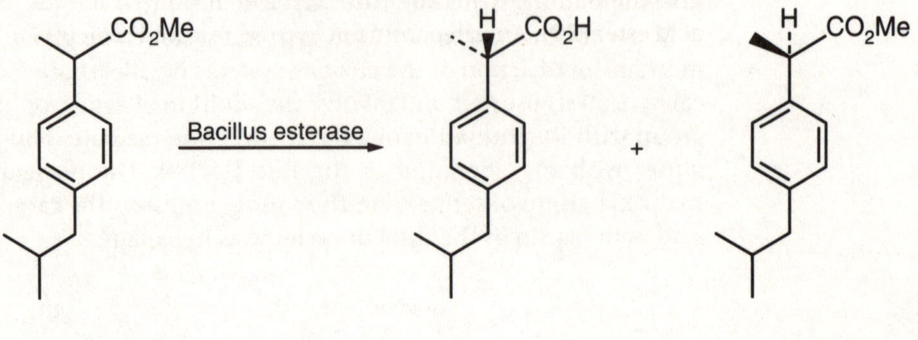
CO2Me
Bacillus esterase
H
CO2H
+
H
CO2Me

Table 2.1 Some enzymes used for hydrolysis

α-Chymotrypsin
Pig liver esterase
Horse liver esterase
Cholesterol esterase
Lyophilized baker's yeast
Penicillin acylase
Porcine pancreatic lipase
Candida lipase
Pseudomonas cepacia lipase
Bacillus carboxyl-esterase
Aspergillus oryzae protease

α-Chymotrypsin has been very widely used in ester and peptide hydrolysis. The mechanism that is proposed involves a catalytic triad of amino acids, serine-195, histidine-57 and aspartate-102 at the catalytic site. The mechanism is shown in Scheme 5. Its essential feature is the dual role that the basic imidazole ring in the catalytic trio plays in enhancing the reactivity of the serine, firstly in the formation of the serine: acyl enzyme intermediate and then secondly, facilitating the hydrolysis of the acylated serine and the release of the acid. During this sequence the positively charged imidazole ring is stabilized by interaction with the negatively charged aspartate-102. As a result of extensive studies, an active site model has been proposed into which the substrate must fit. This model is shown in Figure 2.1. It possesses a hydrophobic pocket, a binding site utilizing a hydrogen bond form *N*-acyl or O–H groups, a small volume site and a site for nucleophilic attack.

Some examples of the application of α-chymotrypsin to ester hydrolysis are as follows. α-Chymotrypsin exhibits an L-stereospecificity towards the hydrolysis of amino acid esters and it is therefore used in the resolution of racemic esters. For example, *N*-acetyl-DL-phenylalanine methyl ester (**27**) is resolved to give the L-amino-acid (**28**), the natural enantiomer, and the unreacted D-amino-acid ester (**29**). Another amino acid to be resolved is diethyl DL-*N*-acetylaspartate

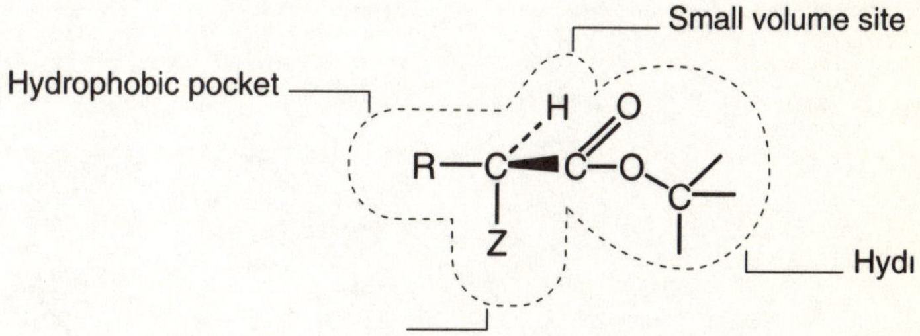

Fig. 2.1 Active site model for α-chymotrypsin.

Asp His Ser

O O⁻ HN N H O C OR² R¹

Scheme 5 The mechanism of ester hydrolysis by α-chymotrypsin.

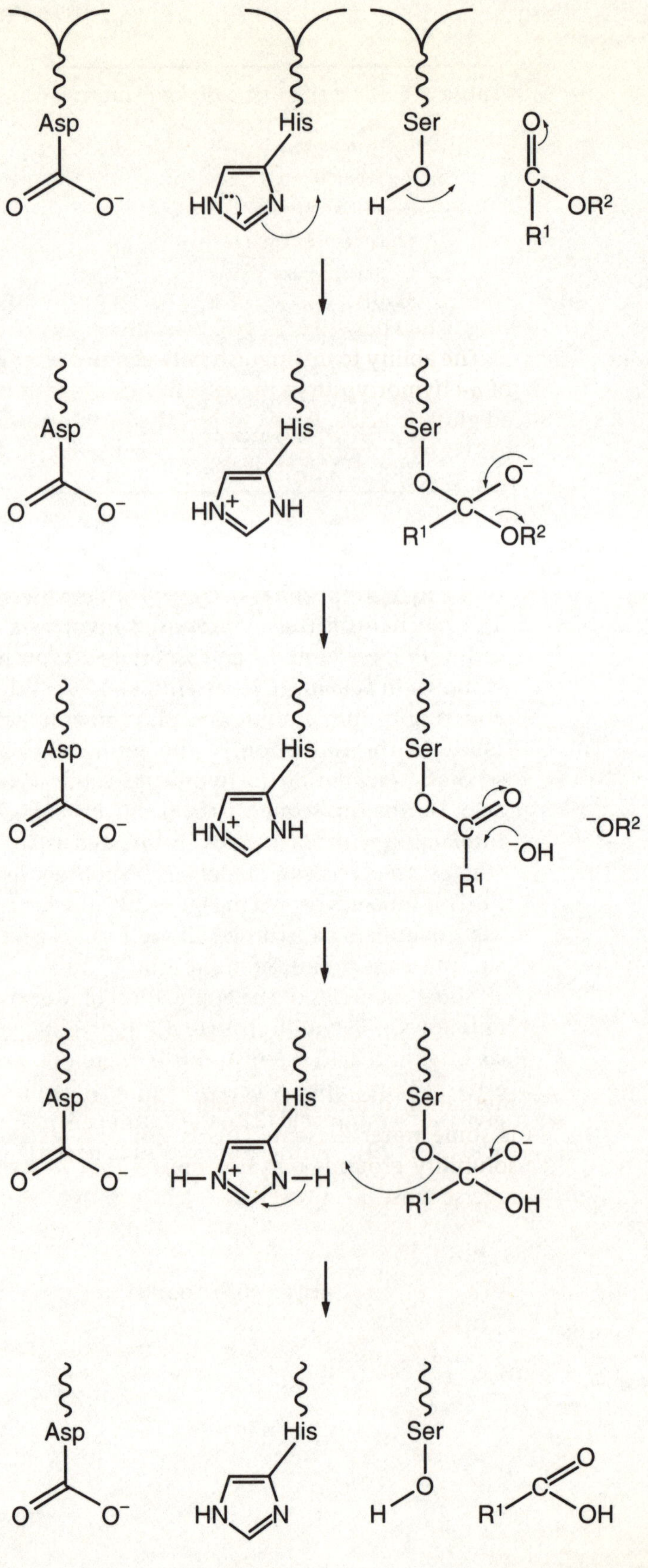

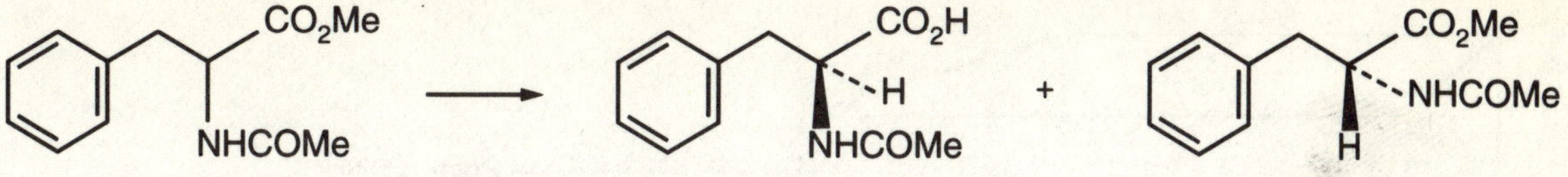

27 28 29

(**30**) where the enzyme system showed not only L-stereospecificity but also distinguished between the two ester groupings giving (**31**) and (**32**).

The ability to distinguish between two ester groupings has permitted the use of α-chymotrypsin in the selective cleavage of the pro-chiral esters of substituted glutaric acids. For example, the pro-*S* ester of dimethyl-β-hydroxyglutarate (**33**) is hydrolysed to give the mono-carboxylic acid (**34**).

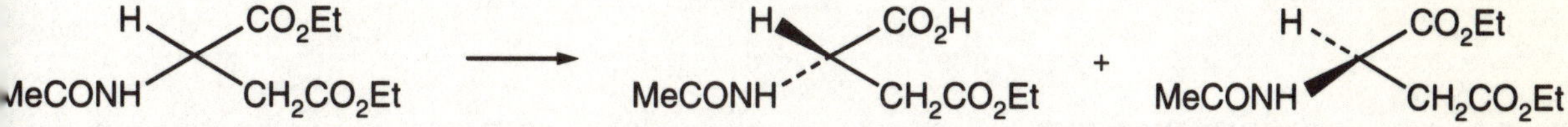

30 31 32

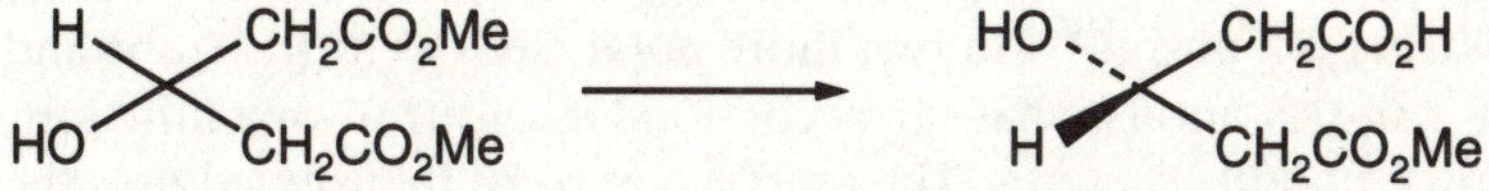

33 34

Although we normally consider enzyme systems operating in an aqueous environment, α-chymotrypsin can be used in organic solvents such as methanol, dimethylsulfoxide and dimethylformamide provided that there is some water present. This is not only extends the range of substrates on solubility grounds but also means that α-chymotrypsin can be used in the reverse sense, i.e. to prepare esters, e.g. to convert (**35**) to (**36**).

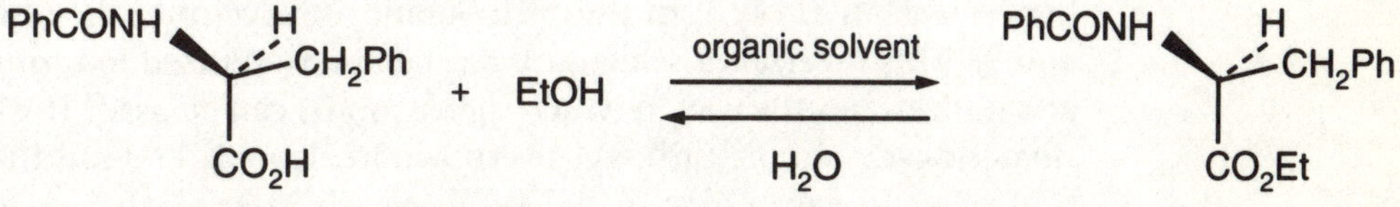

35 36

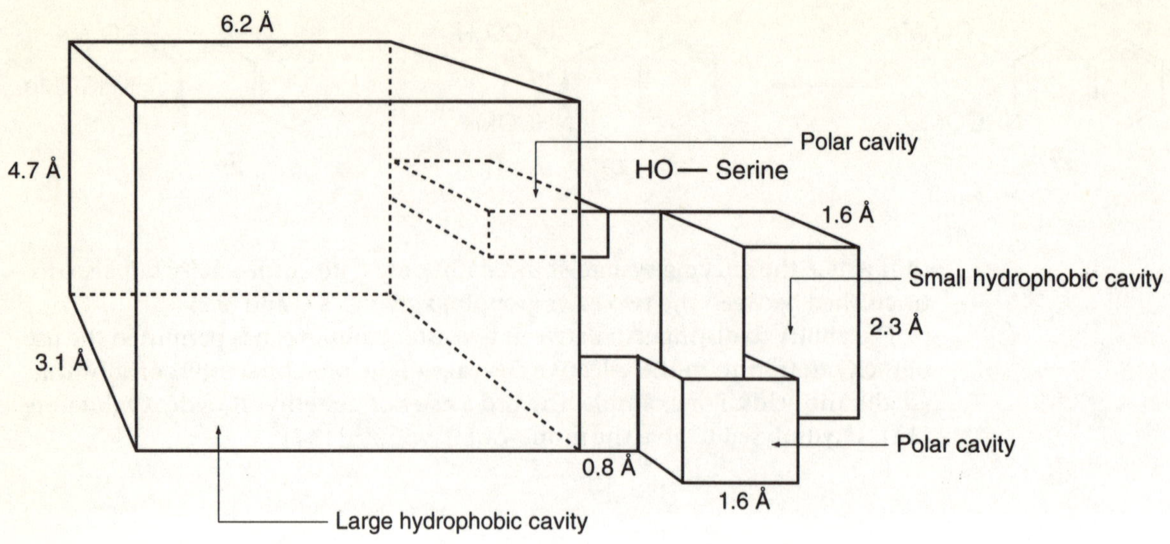

Fig. 2.2 Active site model for pig liver esterase.

Pig liver esterase (PLE) is another readily available systems that has wide applicability. Its first use in asymmetric synthesis was recorded in 1903. Like chymotrypsin it is relatively inexpensive and it has a broad substrate specificity and high stereoselectivity. Efforts have been made by several groups to develop active site models that would permit the interpretation and prediction of this enzyme's stereospecificity. One active site model (Figure 2.2) for PLE has been proposed by J.B. Jones. It contains four binding regions, a small and a large hydrophobic pocket (H_s and H_l) and two more polar cavities at the front and back of the active site (P_F and P_B) that can accommodate electron-rich functions and act as hydrogen bond acceptors. The ester group to be hydrolysed must be placed within the orbit of the serine nucleophile. The boundaries of the pockets represent the constraints that are imposed by the presence of the peptide chain that constitutes the enzyme.

Some examples of the use of PLE are as follows. Its application to the selective hydrolysis of pro-chiral glutaric acid esters has been studied particularly in situations where the stereogenic centre is in the β-position. For example, the selective hydrolysis of the pro-*R* methyl ester group of dimethyl β-hydroxy-β-methyl glutarate (**37**) afforded the half-ester (**38**), which was in turn converted to mevalonolactone (**39**). The hydrolytic resolution of some racemic monocyclic esters using PLE has been examined revealing the high stereoselectivity of this system. In Schemes 6 to 9, there is an interesting switch in stereoselectivity between the hydrolysis of the cyclobutane and cyclohexane esters (Schemes 6 and 8). This reversal of selectivity has been rationalized in terms of the model and it illustrates the way in which such a model can be used. The fit of the enantiomeric cyclobutane carboxylates shown in Figure 2.3 reveals the accommodation of one enantiomer in the H_S space whilst the other enantiomer has a

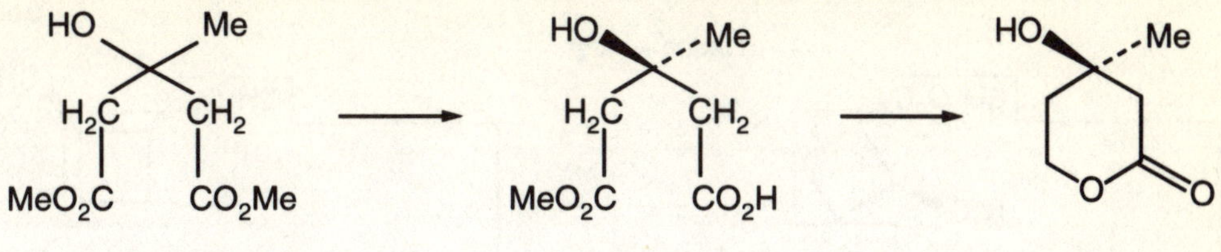

37 → 38 → 39

HO Me
H_2C CH_2
MeO_2C CO_2Me

HO Me
H_2C CH_2
MeO_2C CO_2H

HO Me
O O

Scheme 6

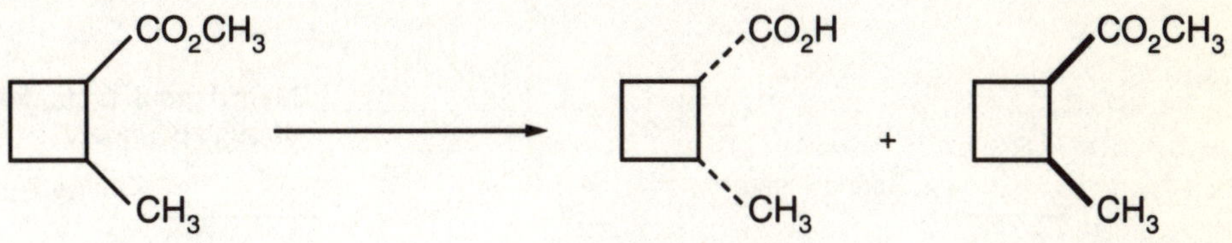

CO_2CH_3, CH_3 (±) → CO_2H, CH_3 90% ≥ 97%ee + CO_2CH_3, CH_3 64% ≥ 97%ee

Scheme 7

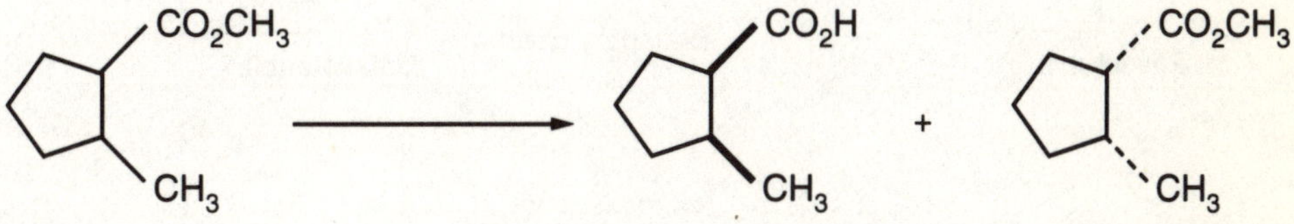

CO_2CH_3, CH_3 (±) → CO_2H, CH_3 57% 22%ee + CO_2CH_3, CH_3 42% 17%ee

Scheme 8

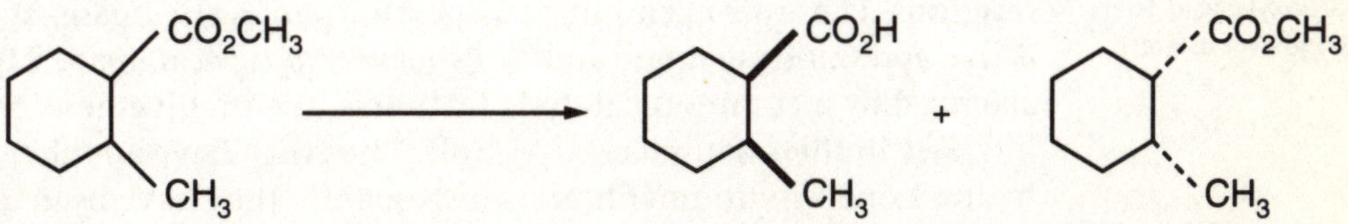

CO_2CH_3, CH_3 (±) → CO_2H, CH_3 93% ≥ 97%ee + CO_2CH_3, CH_3 91% ≥ 97%ee

Scheme 9

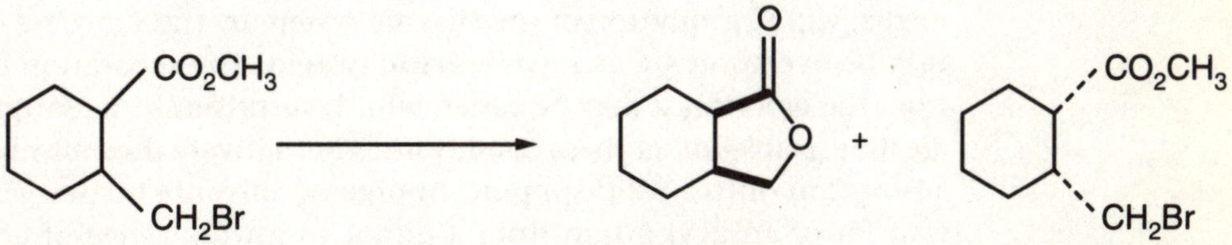

CO_2CH_3, CH_2Br (±) → O, O 98% ≥ 97%ee + CO_2CH_3, CH_2Br 100% ≥ 97%ee

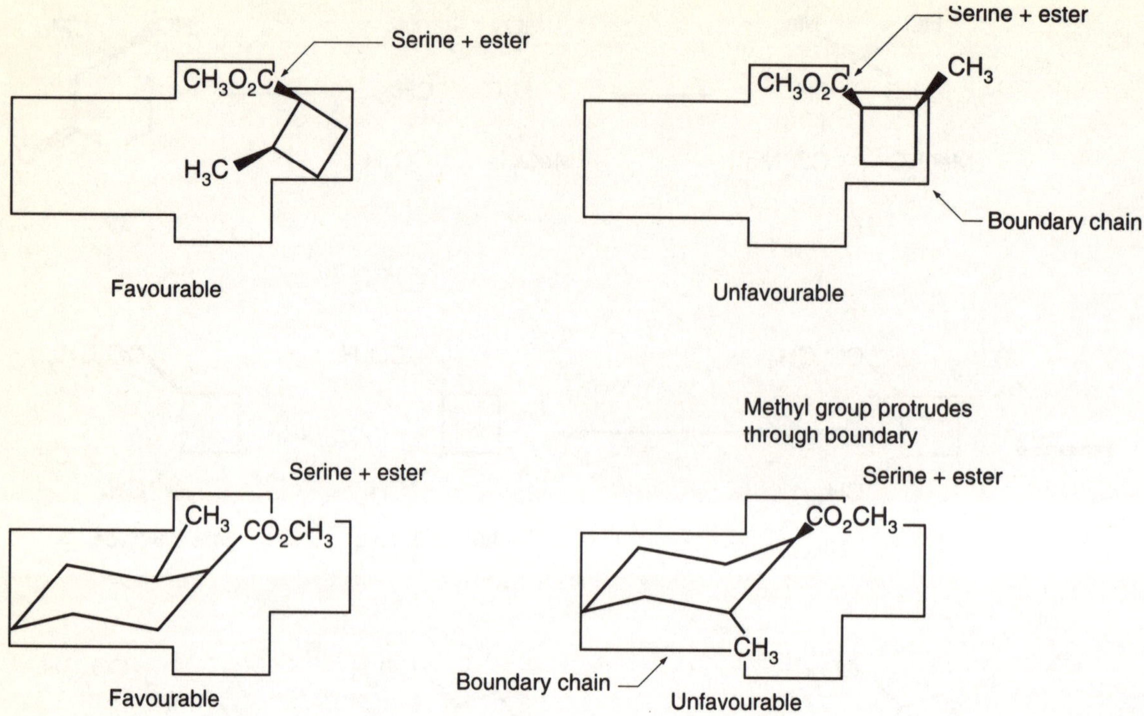

Fig. 2.3 Favourable and unfavourable modes of binding of cyclobutane and cyclohexane carboxylic acid esters to pig liver esterase.

methyl group protruding into the boundary peptide chain. In the cyclohexane case the ring systems occupy the H_L space and again in one enantiomer a methyl group interacts with the boundary.

A number of lipases have found use in both hydrolytic and esterification reactions. These are exemplified by porcine pancreatic lipase, by *Candida cylindracea* and *C. rugosa* lipase and by *Pseudomonas cepacia* lipase. Lipases are characterized by a common catalytic feature — the involvement of a lipid–water interface in their natural catalytic role. Thus they have an inherent affinity for hydrophobic environments and consequently they have been used in organic solvents or biphasic systems. There are a number of advantages in conducting biotransformations under these conditions. These include the use of substrates that are poorly soluble in water, the shift of hydrolytic reactions in the reverse synthetic sense and the suppression of undesirable side reactions in organic media, whilst inhibition of the enzyme system by the substrate or the product may be overcome by using a biphasic system. The separation of the product from the biocatalyst may be easier, whilst, surprisingly, a few enzyme systems are more stable under these conditions. Alternatively they may be stabilized by adsorption onto a solid support. In organic solvents lipases will catalyse the transfer of an acyl group from a donor to a wide range of acceptors other than water, thus facilitating not only esterification reactions but also transesterification steps.

Probably the most widely used lipase is porcine pancreatic lipase (PPL), although the crude preparation is a mixture and as such it is sometimes known as 'pancreatin' or 'steapsin'. Some applications of PPL are as follows. It is sometimes found that different hydrolases show opposite specificity. This is exemplified by the regio- and enantioselective hydrolysis of dimethyl α-methylsuccinate (**40**), in which α-chymotrypsin exhibited the same enantio- but opposite regio-selectivity (see Scheme 10). This experiment also revealed a second point – namely, that the overall enantioselectivity of a particular hydrolysis may often be improved by recycling the product through a second hydrolysis.

CH_3 CO_2CH_3 CO_2H 73% e.e. → 97% e.e. + CH_3 CO_2CH_3 CO_2CH_3 >95% e.e.

PPL

CH_3 CO_2CH_3 CO_2CH_3 **40**

α-Chymotrypsin

CH_3 CO_2H CO_2CH_3 70% e.e. + CH_3 CO_2CH_3 CO_2CH_3 76% e.e.

Scheme 10 Differing selectivities in the enzymatic hydrolysis of dimethyl α-methylsuccinate.

The asymmetric hydrolysis of some cyclic *meso* diacetates such as (**41**) provided some chiral alcohols, which are building blocks of value in prostaglandin synthesis. In other work, hydrolysis of the diacetate ((**42**)R = CO_2Me) by PPL gave the mono-ester ((**42**) R = H) in 96% yield in an essentially optically pure state. This in turn was converted to the azido-alcohol (**43**), which was used for the preparation of nucleoside analogues such as (**44**), which is a potent antiviral agent.

R= OAc OAc → R= OAc OH

41 R = HO H ; H AcO

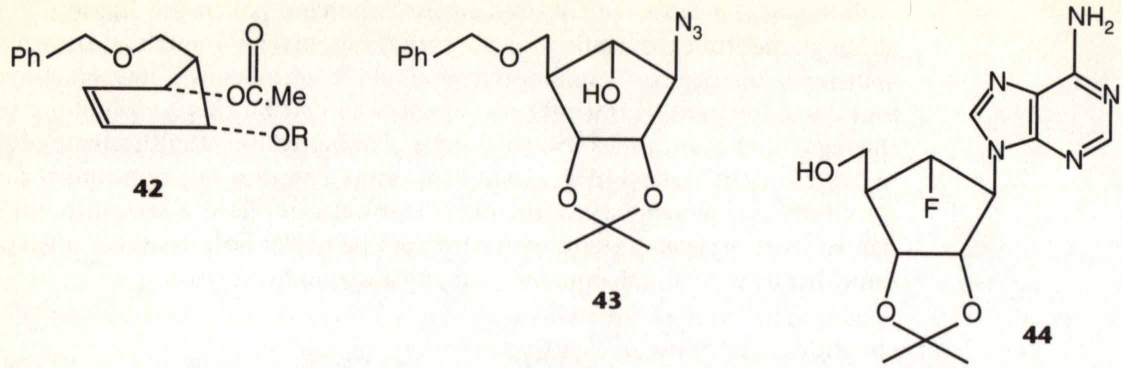

42

43

44

In trans-esterification reactions, enol esters such as vinyl acetate and isopropenyl acetate, can usefully function as acyl donors. An advantage of these sources of an acetate unit, is that the enol that is released rapidly tautomerizes to acetaldehyde or acetone, thus minimizing the reverse reaction. This is a convenient acetylation procedure.

The lipases from the yeast, *Candida cylindracea* (*rugosa*) (CCL) and from the bacterium, *Pseudomonas cepacia* (PCL), have found a number of applications, particularly in the resolution of secondary alcohols. In particular, Candida lipase appears to be able to accommodate relatively bulky esters at its active site. Some examples are shown by the enzymatic resolution of cyclohexane-1,2,3-triol esters (Schemes 11 to 13).

Scheme 11 $OCOC_3H_7$ (±) —PSL→ OH >95%ee + $OCOC_3H_7$ >95%ee

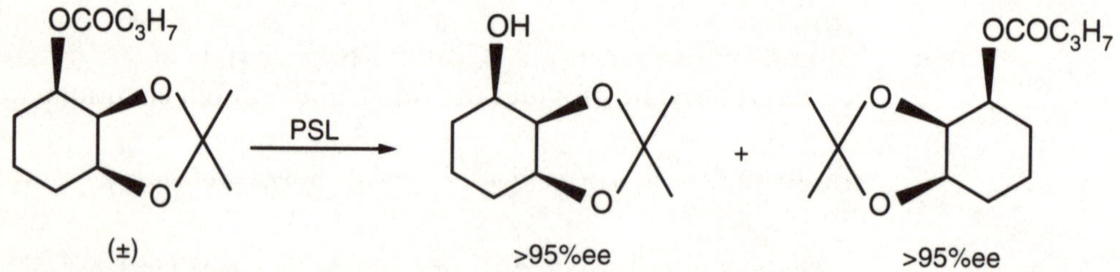

Scheme 12 $OCOC_3H_7$ (±) —CCL→ OH >95%ee + $OCOC_3H_7$ >95%ee

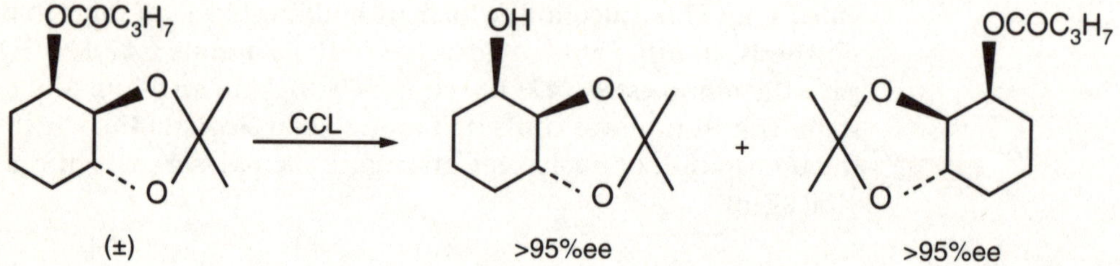

Scheme 13 $OCOC_3H_7$ (±) —PPL/or CSL→ OH >95%ee + $OCOC_3H_7$ >95%ee

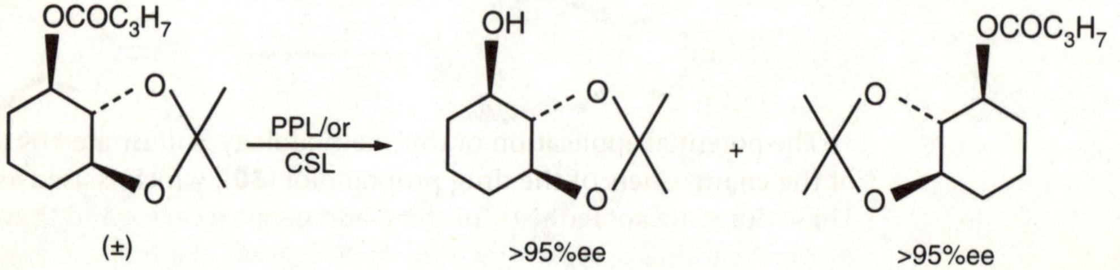

A rule based on the relative sizes of the substituents at the stereocentre has been proposed, which predicts the enantiomer that will react faster in the kinetic resolutions by bovine cholesterol esterase, *Candida rugosa* lipase and *Pseudomonas cepacia* lipase. The stereochemistry of this enantiomer is shown in Figure 2.4. X-Ray crystal structures of the covalent complexes between the *Candida rugosa* lipase and the phosphonate transition state analogues of ester hydrolysis — 1*R* and 1*S* methylhexyl phosphonates — have established that the large methyl unit of the slower reacting enantiomer interacts with the histidine ring of the catalytic triad for this enzyme (Glu.His.Ser.), disrupting the hydrogen bonding involved in the hydrolysis. A common orientation of the loops in the peptide chain of the lipases that assemble the catalytic triad at the active site appears to account for their common enantiopreferences in hydrolysis.

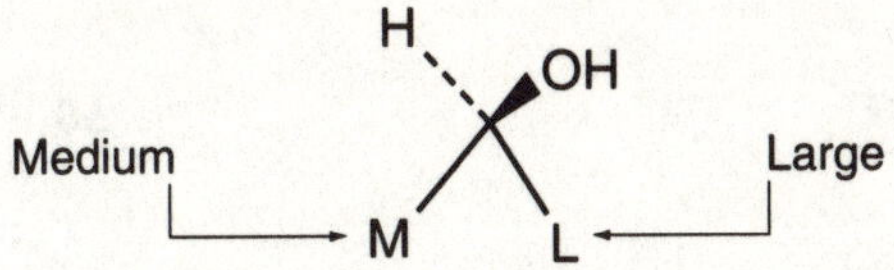

Fig. 2.4 The stereochemistry of the hydrolysis product formed by *Candida rugosa* esterase.

H OH
Medium → M
Large → L

An interesting application of this system is found in the preparation of the lactic acids. Enantiomerically pure *S*-(+)-lactic acid (**45**) is readily available but the other enantiomer, *R*-(–)-lactic acid (**46**) is rarer and more expensive. It can be prepared by the D-lactate dehydrogenase catalysed reduction of pyruvate or by the microbial destruction of the (*S*)-enantiomer in a racemic mixture. However, the PCL catalysed hydrolysis of the tertiary-butyl ester of lactyl butyrate provides the (*R*)-alcohol in high enantiomeric excess, (**47**) to (**48**).

HO H CH_3 CO_2H **45**

H OH CH_3 CO_2H **46**

$OCOC_3H_7$ CH_3 $CO.OC(CH_3)_3$ **47** —PCL→ H OH CH_3 $CO.OC(CH_3)_3$ **48** + $OCOC_3H_7$ H CH_3 $CO.OC(CH_3)_3$

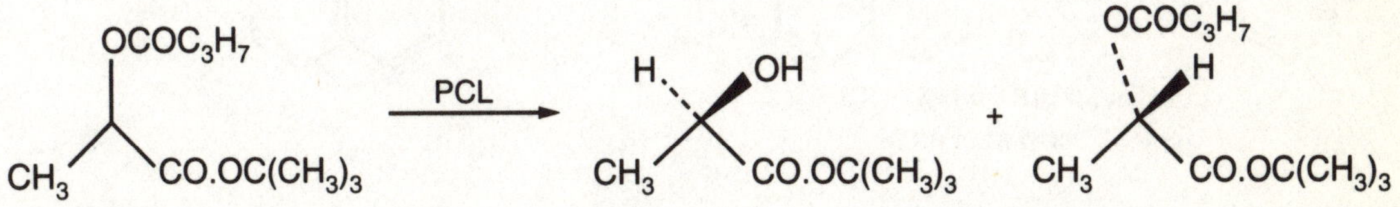

The potential application of this methodology is illustrated by a preparation of the enantiomers of the drug propranolol (**50**), which is used as a β-blocker. These drugs are antagonists for the β-adrenergic receptor and they find medicinal application in the treatment of heart disease. The activity resides in the (*R*)-

isomer. The key step in the synthesis involves the cleavage of the epoxide (**49**) with isopropylamine to afford (**50**). Hence the production of the (*R*) and (*S*) isomers of the epoxide (**49**) was a target. The strategy is shown in Scheme 14. One approach utilized the enantiospecific lipase-catalysed hydrolysis of an acetate, whilst the other was based on the enantiospecific transesterification with vinyl acetate.

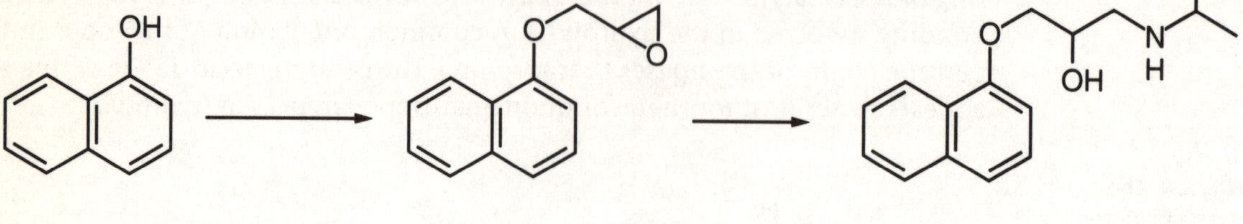

OH

49

50

OAc

Cl

(i)

AcO H

Cl

H OH

Cl

+

S-propranolol

R-propranolol

OH

Cl

(ii)

HO H

Cl

+

H OAc

Cl

(i) PSL, *n*-BuOH, H_2O
(ii) PSL, vinyl acetate

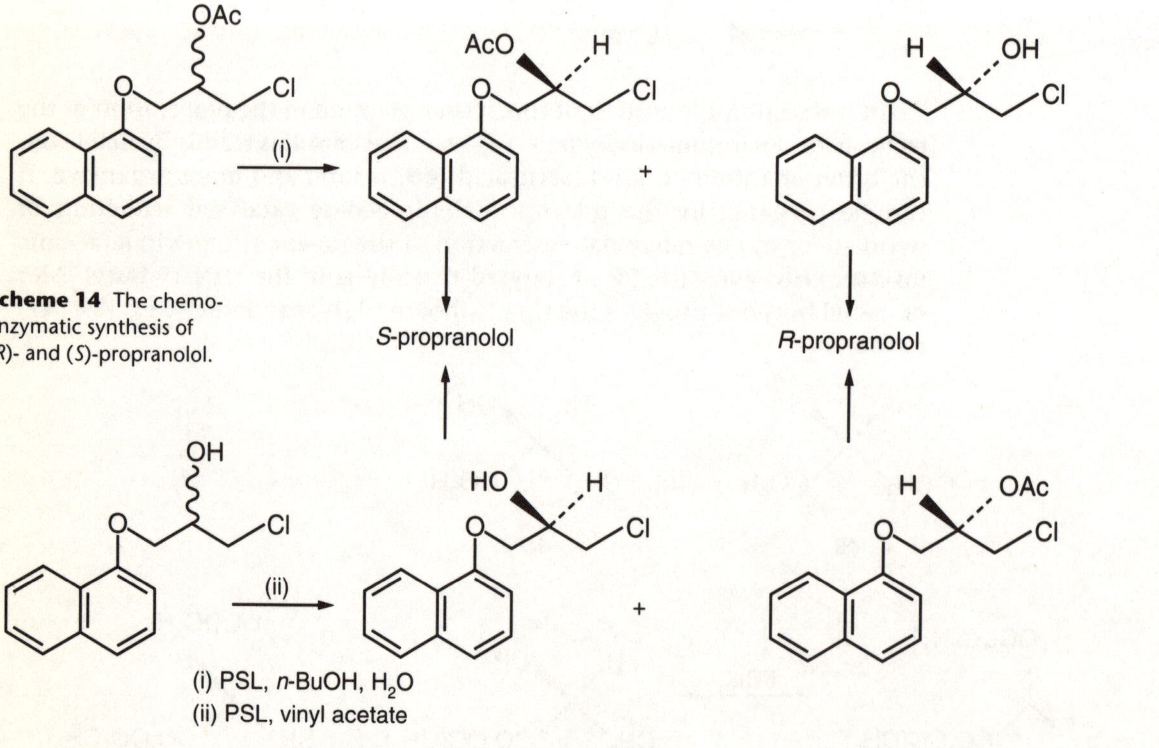

Scheme 14 The chemo-enzymatic synthesis of (*R*)- and (*S*)-propranolol.

Peptide hydrolysis

The use of the specificity of enzymatic hydrolysis for the selective cleavage of peptide bonds has a long history of application in the elucidation of peptide structures. The selectivity of some of the relevant enzyme systems is given in Table 2.2.

Table 2.2 Selectivity of some enzyme systems for peptide hydrolysis.

Enzyme	Selectivity
Trypsin	requires a basic amino acid (arginine or lysine) to provide the >C=O group of the peptide bond
α-Chymotrypsin	requires an aromatic amino acid to provide the C=O of the peptide bond
Pepsin	tends to attack peptide bonds between aromatic amino acids within a peptide chain
Papain	exhibits a relatively broad substrate specificity
Carboxypeptidase-A	hydrolyses the carboxyl terminal peptide bond in a polypeptide particularly if it involves an aromatic amino acid
Clostripain	hydrolyses the carboxyl side of arginine residues
Staphylococcal protease	hydrolyses the carboxyl side of aspartate and glutamate
Elastase	cleaves the peptide bonds between the smaller amino acids

There have been a number of applications of these in the resolution of amino acids. Thus carboxypeptidase-A has been used to hydrolyse the *N*-acetyl derivative of racemic amino acids such as *N*-acetyl-DL-phenylalanine to afford the L-amino acid whilst a related system, hog kidney acylase, has been used in the preparation of labelled amino acids. Most chemical methods for the preparation of 2-deuterio or 2-tritio-amino acids afford the racemate. The subsequent resolution can be achieved using the stereospecificity of the acylase to afford the labelled L-amino acid from the corresponding DL-*N*-acetyl derivative (Scheme 15).

An interesting example of the methodology used in this area comes from work on the production of L-proline – an amino acid that is also a constituent of the antihypertensive drug, captropril (**51**). Enrichment cultures of sewage micro-organisms led to the isolation of a bacterium, *Comamonas testosteroni* which, when grown in the presence of *N*-acetyl-L-proline, produced a specific acylase. Purification of this acylase gave a stable system which rapidly hydrolysed *N*-acetyl-L-proline leaving the *N*-acetyl-D-proline untouched.

Scheme 15

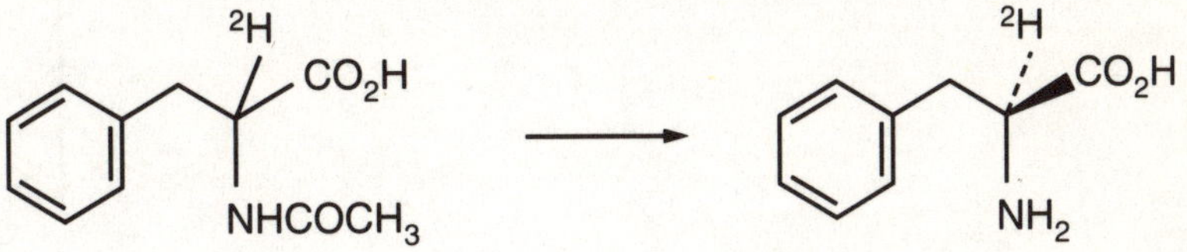

^2H, CO_2H, $NHCOCH_3$ → ^2H, CO_2H, NH_2

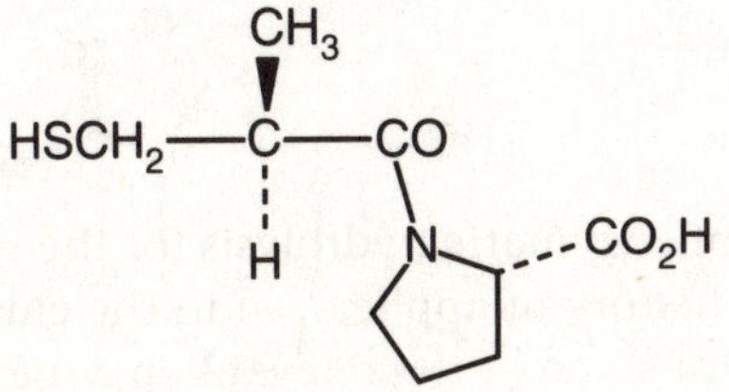

CH_3, $HSCH_2$—C—CO, H, N, CO_2H

51

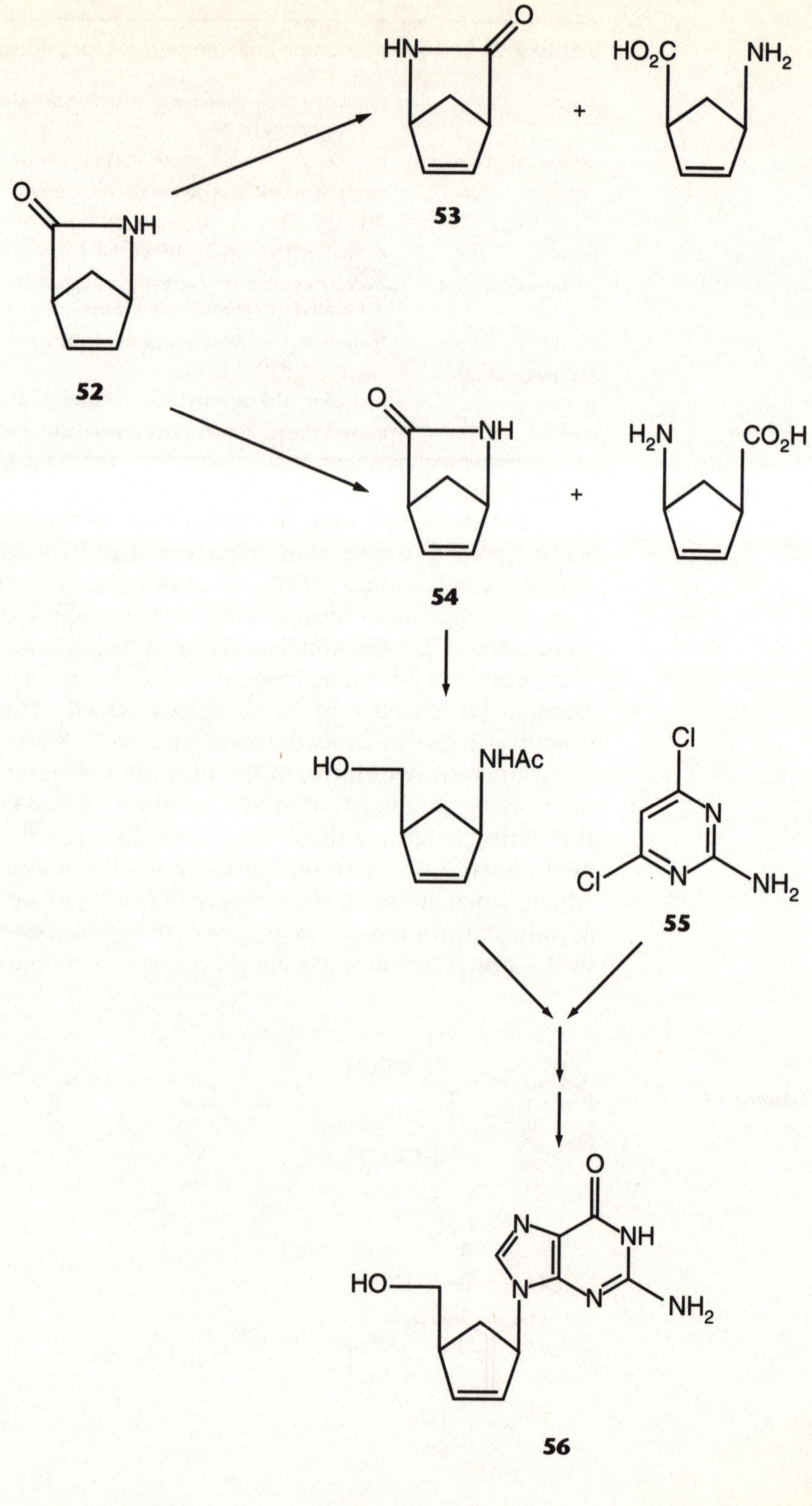

HN O HO$_2$C NH$_2$

+

53

O NH

52

O NH H$_2$N CO$_2$H

+

54

HO NHAc

Cl N Cl N NH$_2$

55

O N NH HO N N NH$_2$

56

More recently, there have been a number of significant applications directed at the selective formation of peptide bonds. These methods often reduce the need for protection and activation and by their very nature avoid the problems of racemization. An example is given in Scheme 16.

Boc.Leu.OEt + H.Arg.N_2H_2Ph —α-Chymotrypsin→ Boc.Leu.Arg.N_2H_2Ph

Boc.Tyr.OEt + H.Gly.N_2H_2Ph —α-Chymotrypsin→ Boc.Tyr.Gly.N_2H_2Ph

Boc.Gly.OH + H.Phe.N_2H_2Ph —Papain→ Boc.Gly.Phe.N_2H_2Ph

Boc.Arg.OCH_3 + H.Ile.N_2H_2Ph —Trypsin→ Boc.Arg.Ile.N_2H_2Ph

Scheme 16 The enzymatic formation of some peptides.

A synthesis of the antiviral carbocyclic nucleoside, (–)-carbovir (**56**) utilizes two organisms with enantiocomplementary activity in hydrolysing the racemic γ-lactam (**52**). One organism, *Rhodococcus equi*, left the (+)-lactam (**53**) untouched whilst the other, *Pseudomonas solanacearum*, left the (–)-lactam (**54**) unhydrolysed. The lactam (**54**) was coupled with 2-amino-4,6-dichloropyrimidine (**55**) and the product was converted to (–)-carbovir (**56**).

Hydratases

A number of enzyme systems mediate the addition of water to unsaturated systems. Fumarase is a well-known enzyme that catalyses the addition of water to fumaric acid (**57**) to form malic acid (**60**) with the stereospecificity shown in (**60**). However, the enzyme does not tolerate much variation in structure. Thus it will not accept the geometric isomer, maleic acid (**61**) and will only transform the fluoro- and chlorofumaric acids (**58** and **59**). The methyl and other halofumaric acids are poor substrates. There is an interesting difference in the

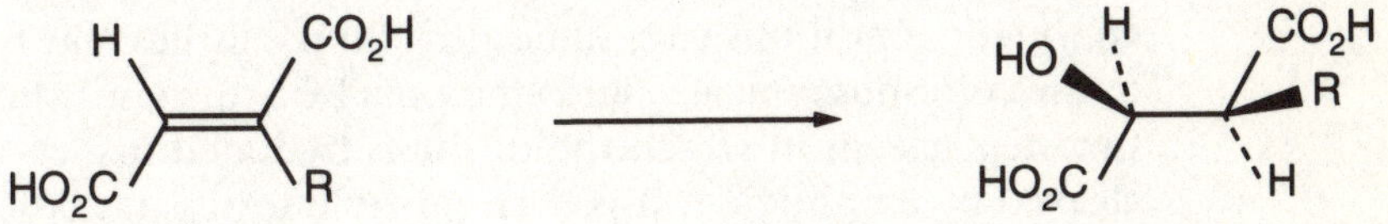

H, CO_2H, HO_2C, R

57 R = H
58 R = F
59 R = Cl

H, CO_2H, HO, R, HO_2C, H

60

H, CO_2H, H, CO_2H

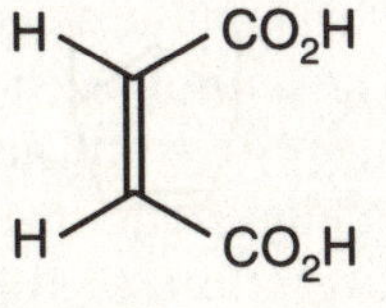

61

regiospecificity of addition to the fluoro- and chlorofumaric acids in that chloromalic acid is obtained whilst the fluorine is lost and oxaloacetic acid is formed. The analogous addition of ammonia catalysed by aspartase and 3-methylaspartase shows a wider structural flexibility.

The addition of water in a *trans* manner to various epoxides is a more widely encountered biotransformation. Indeed this is a common transformation which is observed as a sequel to epoxidation as with arene oxide formation in mammalian liver microsomal systems. The hydrolysis of epoxides is subject to steric factors and proceeds best with terminal and *cis*-1,2-epoxides. The catalytic activity is associated with a histidine residue on the enzyme surface (see Scheme 17).

Scheme 17 The role of histidine in epoxide hydrolysis.

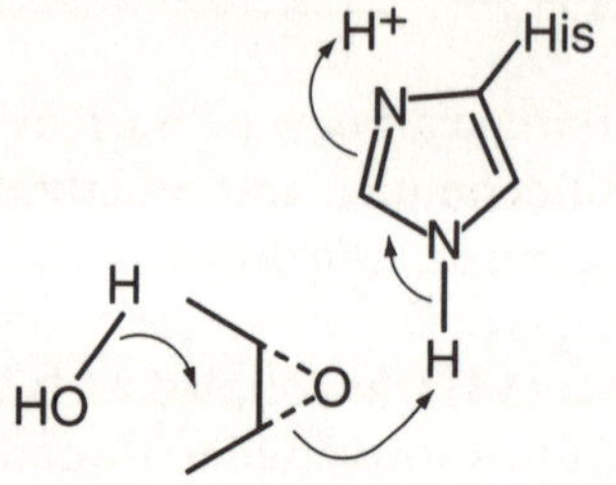

There is some chiral discrimination in the hydrolysis of epoxides by microsomal epoxide hydrolases. However, these hydrolyses may be accompanied by non-enzymatic processes. An interesting example of this has been reported in the dihydroxylation of the remote double bond of geraniol. Epoxidation of the 6,7-double bond of geraniol *N*-phenylcarbamate (**62**) by *Aspergillus niger* affords the 6-(*S*)-epoxide (**63**). This undergoes a spontaneous acid-catalysed hydrolysis at pH 2 to form the 6-(*S*)-diol (**64**). However at pH 6–7, enzymatic hydrolysis occurs to give the 6-(*R*)-diol (**65**).

The hydrolysis of nitriles is a very useful enzymatic process. Chemical methods for the hydrolysis of nitriles often involve quite drastic conditions, such as heating in the presence of strongly basic or acidic catalysts. Consequently, mild enzymatic conditions have some attractions. Nitriles may be converted directly to the corresponding acid and ammonia by a nitrilase. Alternatively, there is a two-step system in which the nitrile is hydrated first to an amide and then cleaved by an amidase. It is then possible to separate the two activities and obtain a system that will convert nitriles to amides.

There is considerable demand for acrylamide (**67**), which is a monomer used in the production of synthetic fibres. It is made by the catalytic hydrolysis of acrylonitrile (**66**). A number of bacteria will selectively carry out this biotransformation. Two organisms, *Pseudomonas chlororaphis* and a Brevibacterium species have been isolated, from which suitable strains have been developed for the biotransformation. A strain of *P. chlororaphis* was obtained that did not produce polysaccharides and which could be immobilized on a polyacrylamide gel. An interesting feature is that the enzyme system utilizes ferric iron and a second prosthethic group, the pyrroloquinoline quinone (PQQ) (**68**), suggesting that the mechanism may involve a redox reaction, possibly based on the nucleophilic hydroperoxide anion.

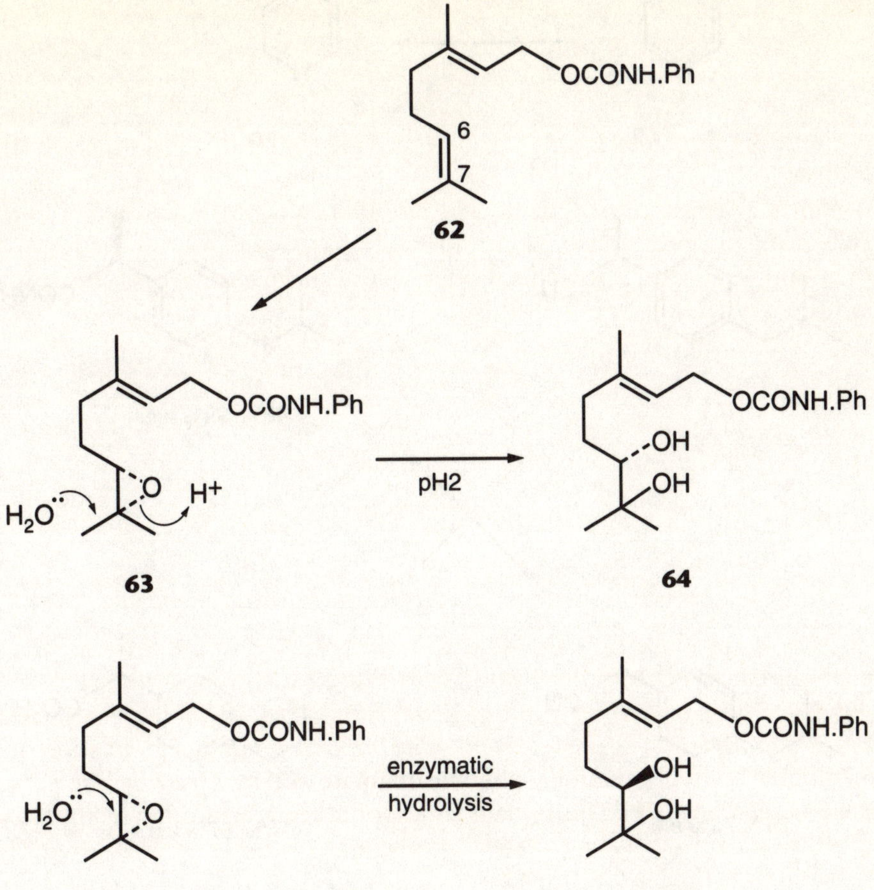

OCONH.Ph
6
7
62

OCONH.Ph
H_2O
O
H^+
63

pH2

OCONH.Ph
OH
OH
64

OCONH.Ph
H_2O
O

enzymatic
hydrolysis

OCONH.Ph
OH
OH
65

$CH_2{=}CH.CN$ → $CH_2{=}CH.CONH_2$

66 67

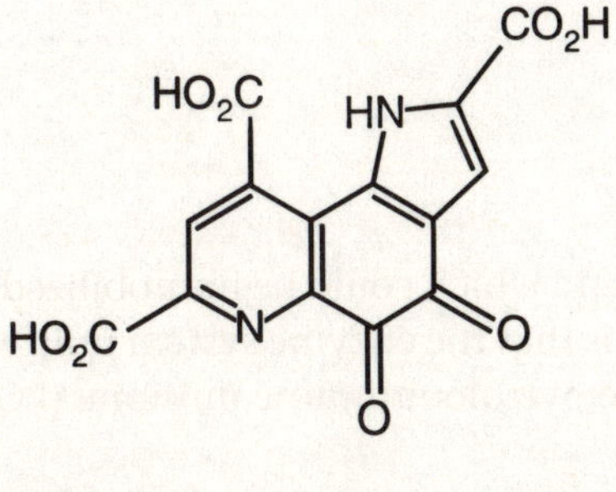

CO_2H
HO_2C
HN
HO_2C
N
O
O
68

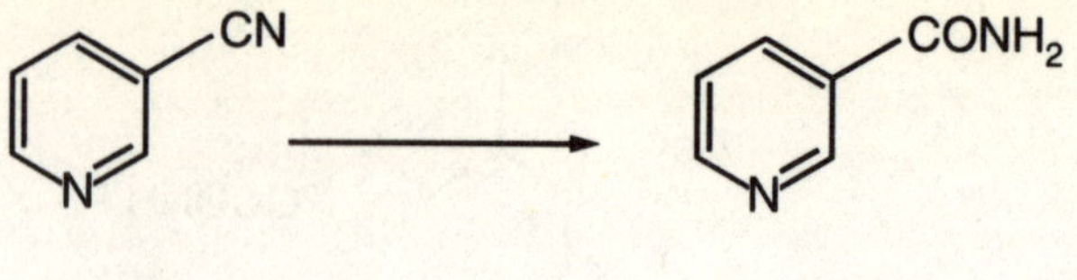

CN CONH$_2$

69 70

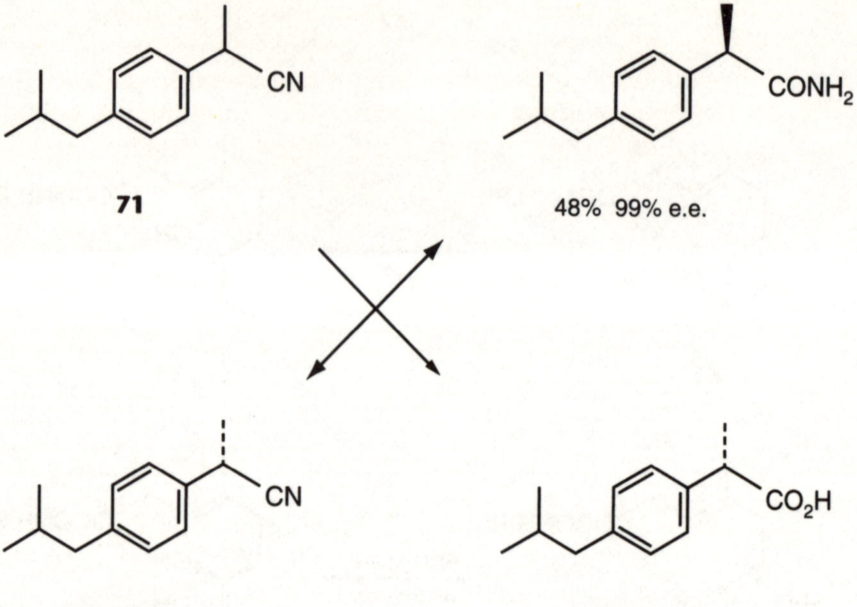

CN CONH$_2$

71 48% 99% e.e.

CN CO_2H

21% 73% e.e. 13% 87% e.e.

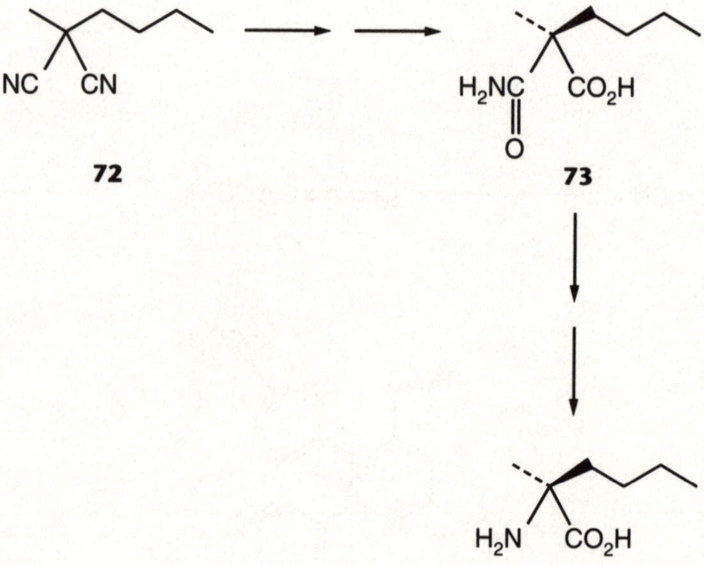

NC CN H_2NC CO_2H O

72 73

H_2N CO_2H

74

This enzyme system will hydrolyse a number of aliphatic nitriles. Hydrolysis of aromatic nitriles to amides employs another organism, *Rhodococcus rhodochrous*, which has quite a broad substrates specificity. This system requires cobalt ions rather than ferric ions and it has been used, for example, to produce nicotinamide (**70**) from 3-cyanopyridine (**69**). Another potentially important application of the hydrolysis of nitriles is exemplified by the use of *Rhodococcus butanica*. This organism will hydrolyse aromatic nitriles to amides and acids. In particular, it will hydrolyse racemic α-arylpropionitriles such as (**71**) with some chiral discrimination. It is possible that the organism has both pathways for nitrile hydrolysis in operation and each has a different enantioselectivity. The arylalkanoic acids (such as ibuprofen) that are produced are important non-steroidal anti-inflammatory agents. Another nitrilase from a *Rhodococcus* species has been immobilized and shown to convert a wide range of nitriles directly to the corresponding acids, thus extending the scope of this methodology. The asymmetric hydrolysis of a disubstituted malononitrile (**72**) by *R. rhodochrous* has been reported. Since the Hofmann rearrangement of amides with an adjacent quaternary asymmetric carbon atom proceeds with retention of configuration, this provides access to amino acids. Thus (**73**) was converted to α-methylnor-leucine (**74**).

3 The formation of C–C bonds

In a synthesis, the formation of a new C–C bond often results in the creation of a new chiral centre. Enantiospecific methods of achieving this are of considerable importance in the synthesis of biologically active molecules. A number of useful synthetic reactions involving the formation of carbon – carbon bonds are mediated by biological systems and proceed in an enantiospecific manner. These form the subject of this chapter.

Biological acyloin condensations

One of the earliest biotransformations, first observed in 1913 and clarified in 1921, was the acyloin condensation of furfural (**75**) with an acetate unit (see below) to form the ketol (**76**) in the presence of yeast. The reaction was

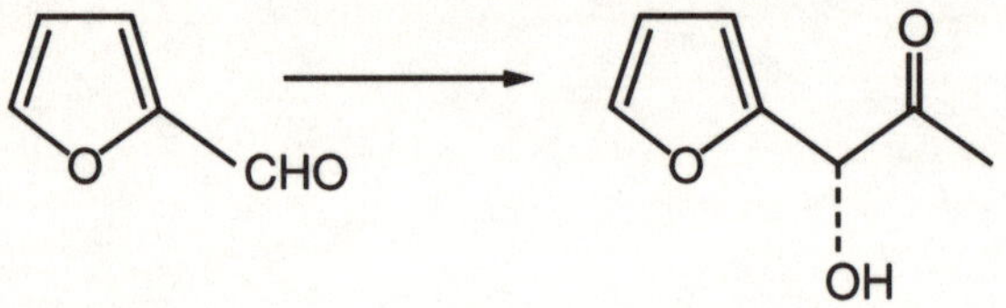

CHO

OH

75 **76**

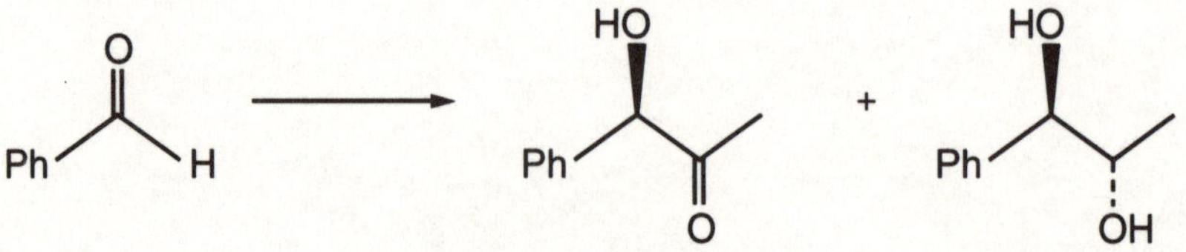

HO HO

O

Ph H Ph Ph

O OH

77 **78**

HO

Ph

NHMe

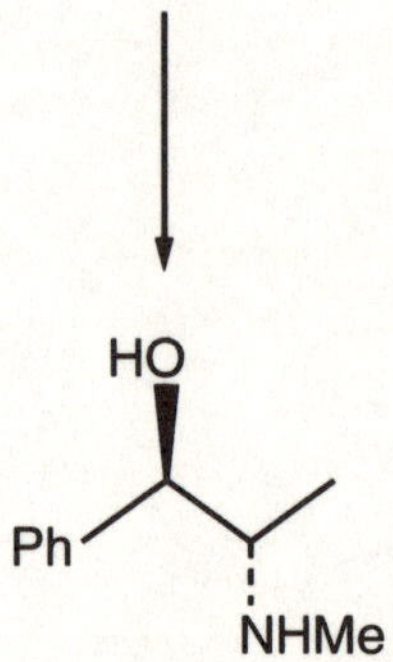

Ph.CH=CH.CHO

79 **80**

subsequently extended using benzaldehyde (**77**) as a substrate to afford an industrial preparation of the alkaloid, (–)-ephedrine (**79**). The acyloin condensation of benzaldehyde and an acetate equivalent mediated by yeast afforded the ketol (**78**) with the stereochemistry shown. Ephedrine (**79**) along with some pseudoephedrine was obtained from the ketol by reaction with methylamine and hydrogenation.

The acyloin reaction utilizes pyruvate decarboxylase as the source of the acetate unit and involves magnesium ions and thiamine pyrophosphate. The mechanism of the reaction is shown in Scheme 18. The pyruvate decarboxylase can accept other unnatural α-keto-acids whilst the condensation will proceed with a range of aldehydes including $\alpha\beta$-unsaturated aldehydes such as cinnamaldehyde (**80**). Although the enantiomeric excesses are high, the yields in these processes are often modest and some aldehydes are inhibitors of the enzyme. Some efforts have been made to optimize this biotransformation.

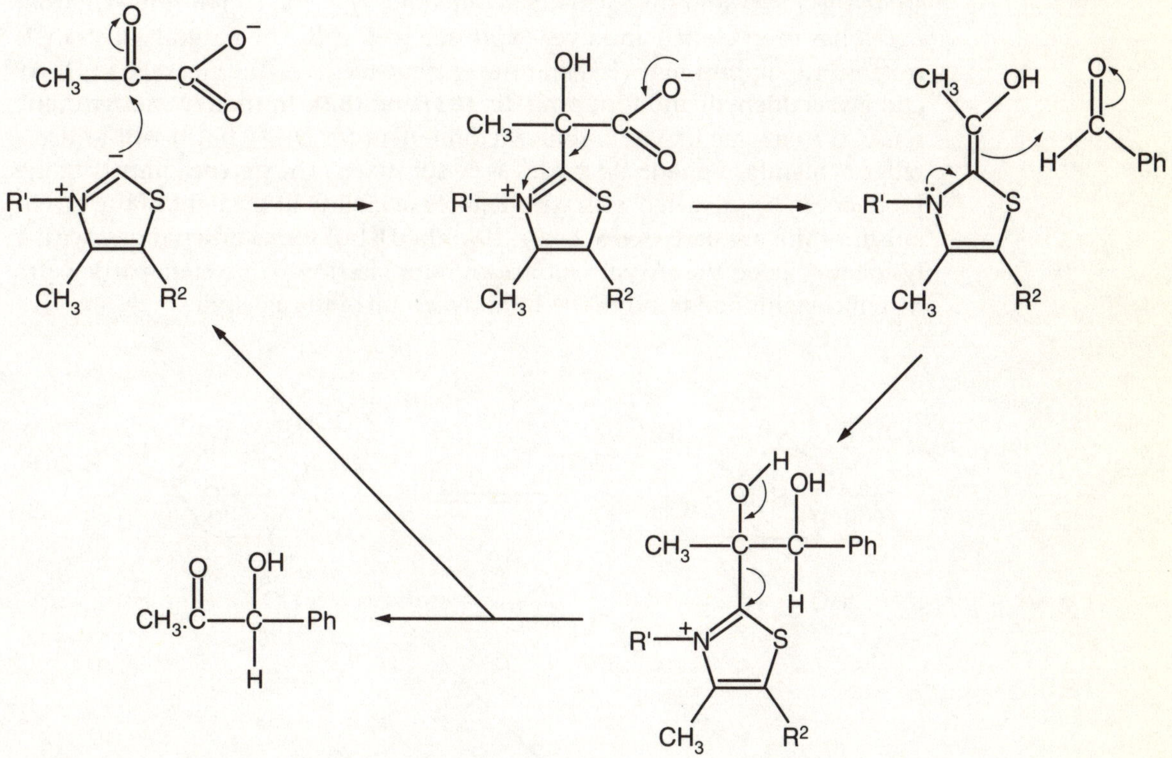

Scheme 18 Thiamine catalysis in pyruvate decarboxylation and the acyloin condensation.

Biological aldol condensations

A number of aldolases have been identified and used in a synthetic sense. These fall into three main groups, based on the structures of the nucleophilic substrates. One group utilizes dihydroxyacetone monophosphate (DHAP), the second phosphoenolpyruvate (PEP) and the third, acetaldehyde. In each case the stereochemistry of the C–C bond that is formed, is controlled by the specific enzyme system involved.

The glycolytic pathway involves retro-aldol reactions and consequently the enzymes of carbohydrate metabolism have provided the source of a number of useful biocatalysts. The aldolase from rabbit muscle, and a similar system from *E. coli*, are readily available and have been used in a number of studies. The aldolases from animals and higher plants differ from bacteria in that the latter have zinc ions at the active site and require potassium ions for activity. Nevertheless, the zinc-containing fructose diphosphate (FDP) aldolase from *E. coli* and the rabbit muscle FDP aldolase appear to have the same substrate specificity and stereochemical outcome. The zinc-containing FDP from *E. coli* has been cloned and over-expressed in *E. coli*. The natural substrate is fructose 1,6-diphosphate (**81**) and the enzyme mediates the cleavage to DHAP and glyceraldehyde monophosphate, (**83**) and (**82**). In the reverse synthetic sense, the enzyme has an absolute requirement for DHAP but it will utilize a range of mainly aliphatic aldehydes as co-substrates. The stereochemistry of the product can be predicted with considerable reliability in the light of the mechanism of the aldolase (see Scheme 19). The DHAP forms an enamine with a lysine residue on the enzyme surface and this electron-rich system participates in a nucleophilic addition to the carbonyl group of the aldehyde.

81

Ⓟ$OCH_2.CH(OH).CHO$ + $HOCH_2.C(=O).CH_2O$Ⓟ   Ⓟ = PO_3^{2-}

82   **83**

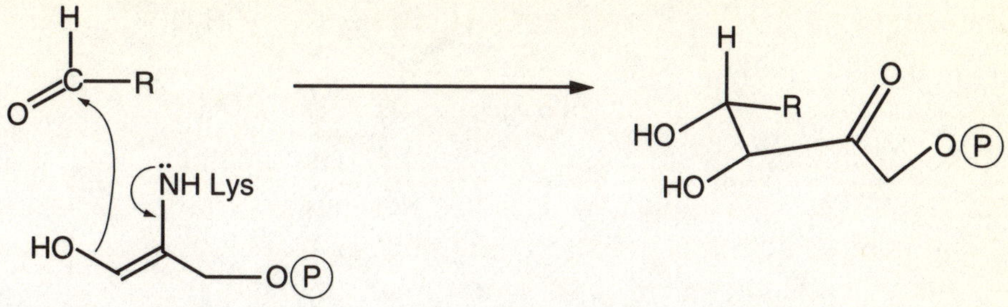

Scheme 19 The role of an enamine in the aldolase condensation.

FDP-aldolase has been widely exploited in the synthesis of unusual carbohydrates, particularly in the work of Whitesides and Wong. A simple synthesis of 5-deoxyfructose is an example, involving the conversion of (**84**) to (**85**). In order to facilitate the isolation of the product, a phosphatase is often added, as in this example, after the aldolase reaction. This preparation proceeded in 67–72% yield, whereas the six-step chemical synthesis starting from fructose was reported to have an overall 13% yield.

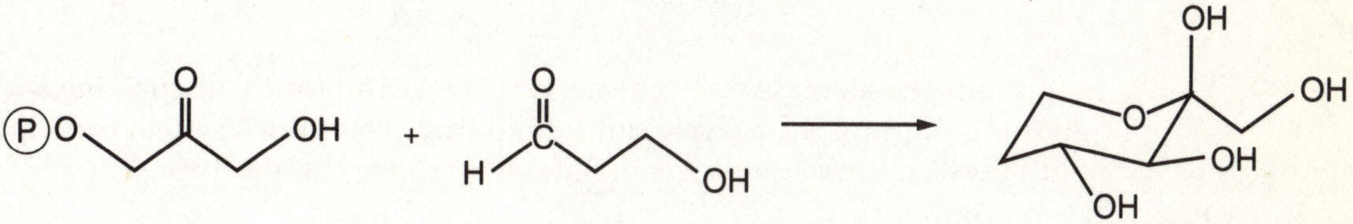

84 **85**

An interesting feature of this system is that it will accept DL-glyceraldehyde monophosphate to give an equimolar mixture of fructose (**81**) and sorbose (**86**) 1,6-diphosphates. The system has also been exploited for the synthesis of labelled sugars (e.g. 2-^{13}C-and 2,5-^{13}C-glucose) via the corresponding labelled fructose derivative. Improved methods for the preparation of DHAP have made this substrate more readily available. It can, for example, be generated enzymatically *in situ* from D-fructose 1,6-diphosphate by using a combination of FDP-aldolase and triosephosphate isomerase. This generates two moles of DHAP from the FDP. A chemical method is based on phosphorylation of the dimeric ethoxy-acetal (**87**) obtained from dihydroxyacetone and triethylorthoformate.

Four complementary aldolases are known in carbohydrate metabolism. Their stereospecificity is shown in Figure 3.1. The relevant enzyme systems have been isolated from strains of *E. coli* that are able to grow on the individual sugars as the sole sources of carbon. Exploratory studies with the purified enzymes of

86

$(EtO)_3CH$, H^+

$(i)\ POCl_3$
$(ii)\ NaHCO_3$

87

H^+

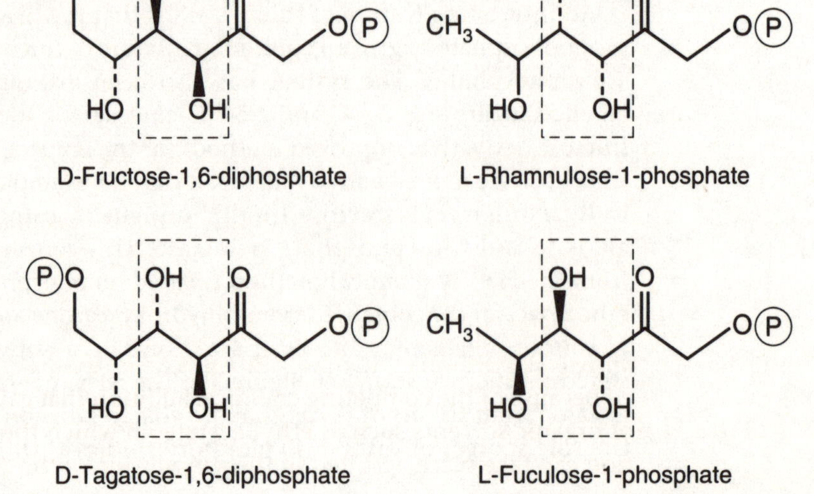

Fig. 3.1 The stereospecificity of the aldolases of carbohydrate metabolism.

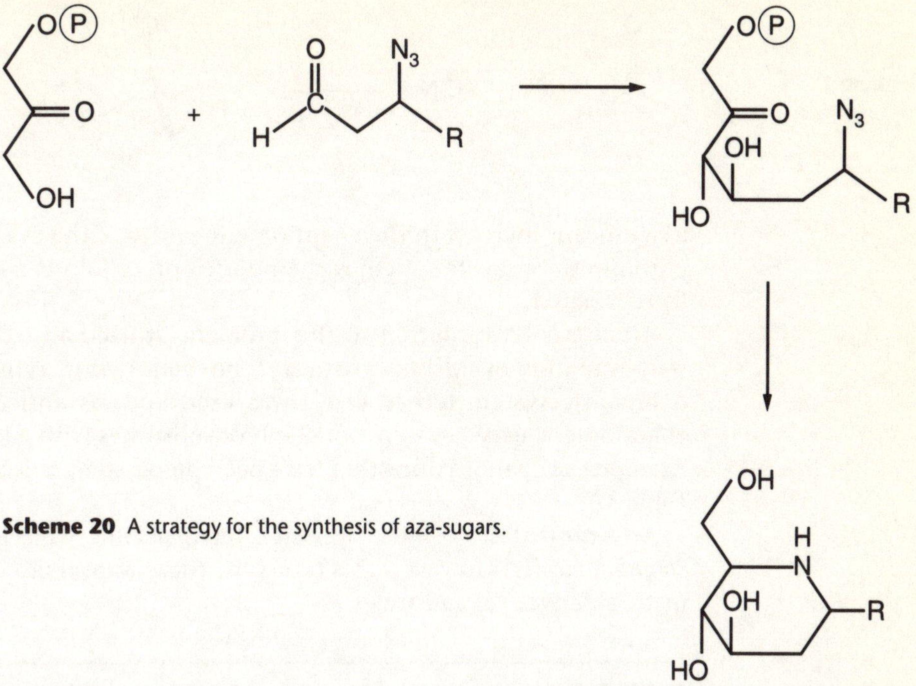

Scheme 20 A strategy for the synthesis of aza-sugars.

rhamnulose aldolase (RhuA) and fuculose aldolase indicated that they had some flexibility in the aldehyde component. However, the diastereoisomeric selectivity in generating the vicinal diol was not complete in the absence of a group in the aldehyde that was capable of hydrogen bonding to the enzyme surface. Nevertheless, the isolation of this group of enzymes provides the basis for a very powerful synthetic methodology for creating chiral polyols of predictable absolute stereochemistry.

The replacement of the ring oxygen of some pyranoses and furanoses by an imino group gives aza-sugars and their analogues, which are of interest as glycosidase inhibitors. A strategy for the asymmetric synthesis of these has been developed utilizing FDP-aldolase and azido-aldehydes as the substrate. This synthetic scheme is outline in Scheme 20.

Cyanohydrin formation and hydrolysis

The enantioselective addition of hydrogen cyanide to benzaldehyde and numerous other aldehydes in the presence of the enzyme mandelonitrile lyase ((*R*)-oxynitrilase) affords optically active (*R*)-cyanohydrins (see Scheme 21).

The requisite enzyme system is readily available from bitter almonds (*Prunus amygdalus*). However, the preparative value of the reaction was originally limited because of the competing chemical addition that gave the racemate. The use of organic solvents such as ethyl acetate, in which the enzyme-catalysed reaction can take place, resulted in the suppression of the chemical reaction and in

Scheme 21

$$RCHO + HCN \longrightarrow R\text{-}CH(OH)\text{-}CN$$

a significant increase in the enantiomeric purity of the cyanohydrin. For these purposes the enzyme system was supported on cellulose, allowing it to be readily re-isolated.

An alternative solution to this problem utilized an oxynitrilase-catalysed transcyanation of aldehydes with acetone cyanohydrin as the source of cyanide. A biphasic system (ether:water) was used and, as anticipated, the highest enantiomeric excesses were obtained for substrates with a low water solubility. Examples of cyanohydrins that have been prepared by this method are given in Table 3.1.

An oxynitrilase with the opposite stereospecificity has been obtained from *Sorghum bicolor*. However, this (*S*)-oxynitrilase appears to be restricted to aromatic aldehydes as substrates.

Table 3.1 Example of Cyanohydrins prepared by oxynitrilase – catalysed transcyanation of aldehydes

Aldehyde	Cyanohydrin	% yield	% e.e.
C_6H_5CHO	HO, H; C_6H_5, CN	72	92
2-$CH_3O.C_6H_4$.CHO	HO, H; $2CH_3OC_6H_4$, CN	65	96
$(CH_3)_3$.C.CHO	HO, H; $(CH_3)_3C$, CN	58	92
CHO (geranial structure)	HO, H; R, CN	46	99

The acid-catalysed hydrolysis of cyanohydrins to α-hydroxy-acids is one of their applications in synthesis. Although in some instances this can be carried out with concentrated hydrochloric acid with minimal racemization, the use of enzymatic hydrolysis of the nitriles is an advantage.

4 Redox reactions

The reduction of a carbonyl group, particularly to generate a new chiral centre, is one of the most widely used biotransformations. The enzymatic methods of reduction take place under mild conditions, minimizing problems from side reactions and with a regio- and stereochemical homogeneity that can be difficult to achieve chemically. Furthermore, these enzymatic methods possess an enantioselectivity that in many cases can be predicted on the basis of empirical rules.

Both intact organisms and isolated enzyme systems have been used. These biological reductions are dependent on the nicotinamide co-enzymes NADH and NADPH, (**88**) and (**89**), as the source of the hydride ion (Scheme 22). If the isolated enzyme systems are to be used catalytically then the co-factors have to be regenerated. Co-factor recycling is normally achieved by coupling the redox process to a second enzyme-catalysed process. There are two common methods for doing this. In the first, formate dehydrogenase is used to catalyse the

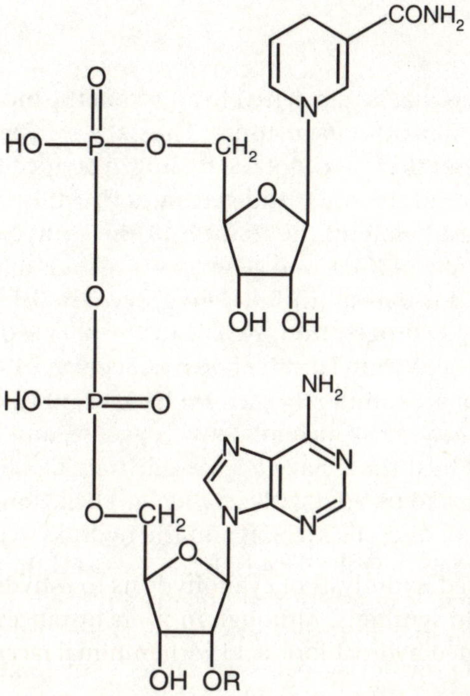

88 R = H

89 R = PO_3H_2

Scheme 22

oxidation of formate to carbon dioxide and thereby regenerate reduced NADH as in Scheme 23. Another method utilizes the oxidation of glucose catalysed by glucose dehydrogenase which, unlike formate dehydrogenase, can be used for both NAD^+ and $NADP^+$ A variant on this uses the oxidation of glucose-6-phosphate by glucose-6-phosphate dehydrogenase. These methods have been adapted to provide deuteriated samples of the co-factors.

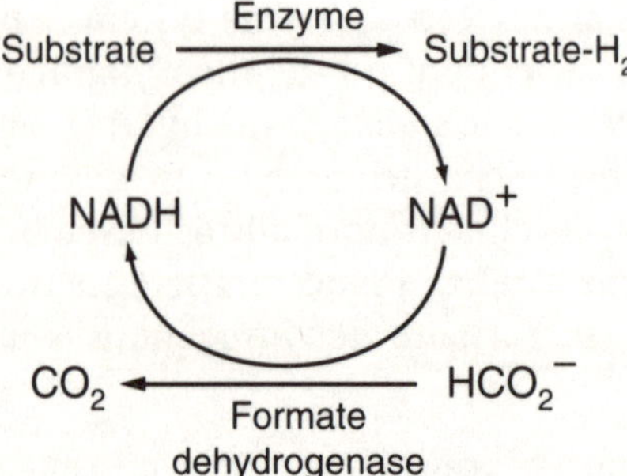

Scheme 23 Co-factor recycling with formate dehydrogenase.

The C–4 hydrogen that is transferred to and from the nicotinamide (Scheme 22) is removed in a stereospecific manner. Thus, the enzyme formate dehydrogenase (FDH) catalyses the transfer of the hydrogen bonded to the carbon of the formate to the *re* face of the NAD^+ and generates H_R whilst the glucose-6-phosphate dehydrogenase catalyses the transfer of the C_1 hydrogen of glucose-6-phosphate to the *si* face of NAD^+ and generates H_S. This is important in labelling the nicotinamide. Horse liver alcohol dehydrogenase (HLADH) catalyses the transfer of the 4-H_R hydrogen from NADH to the *re* face of an aldehyde substrate. The fate of a deuterium label is shown in Scheme 24.

The enzymes that are commonly used are HLADH, yeast alcohol dehydrogenase, *Pseudomonas testosteroni* alcohol dehydrogenase, and the hydroxysteroid dehydrogenases. Whilst these have a wide substrate flexibility, yeast alcohol dehydrogenase tends to be used with smaller acyclic ketones while HLADH is used with mono- and bi-cyclic ketones and the hydroxysteroid dehydrogenases, as their name suggests, with steroids. Baker's yeast is the commonest microorganism that is used. When an intact organism is used, it is not necessary to provide a separate system to regenerate the co-factor. The metabolism of the glucose by the yeast provides the necessary source of hydrogen. This operational advantage is tempered by the fact that baker's yeast contains a complex mixture

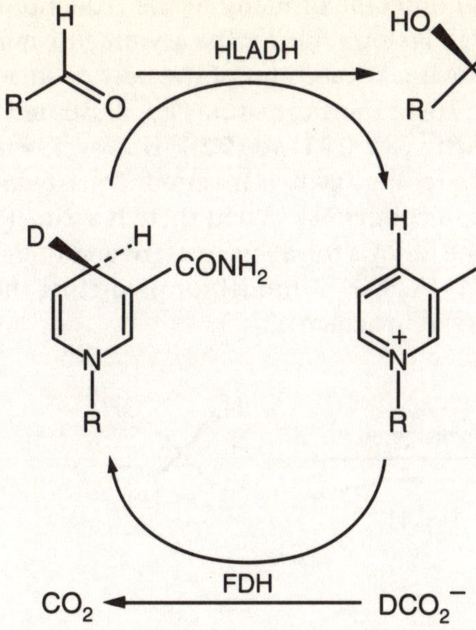

Scheme 24 The stereochemistry of labelling in the HLADH reduction of an aldehyde.

of enzymes including several dehydrogenases of differing stereospecificity. The use of the various systems is illustrated by the following examples.

Reduction of aldehydes and acyclic ketones

The stereochemistry of a number of stages in terpenoid and steroid biosynthesis was established by the use of stereospecifically labelled mevalonates (**90**). The label at the 5-position was introduced enzymatically as shown in Scheme 25.

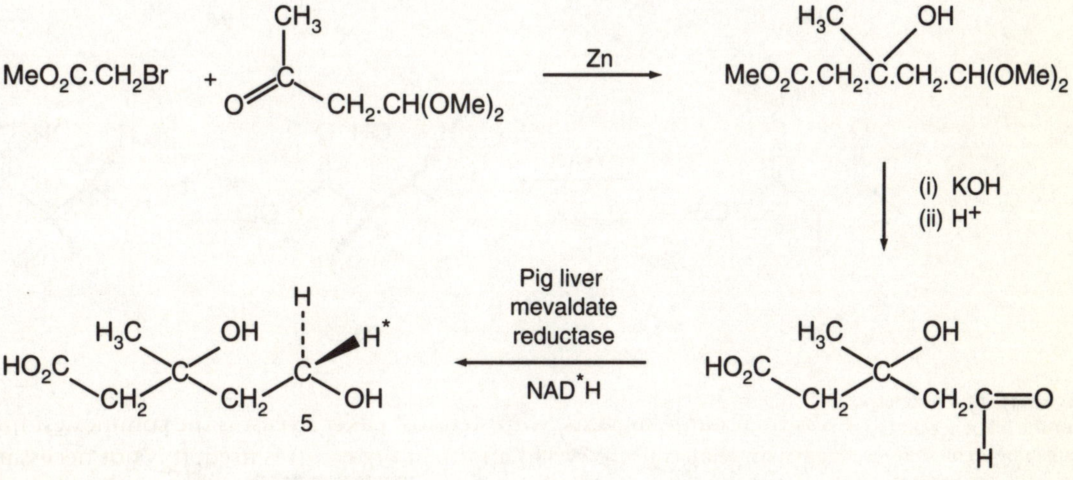

Scheme 25 The preparation of 5-(*R*)-labelled mevalonolactone.

The stereochemical outcome of many of the reductions follows a pattern, which is summarized in Prelog's rule for the asymmetric microbial reduction of ketones. This rule, which takes account of the bulk of adjacent substituents, is shown in Scheme 26. These transformations are illustrated by the reduction of ethyl acetoacetate with yeast [(**91**) to (**92**)]. However, when the alkyl chain becomes longer, the stereospecificity is inverted. Thus, reduction of the β-keto-ester (**93**) gave the (*R*)-alcohol (**94**). When there is a 2-methyl substituent, the reduction of the ketone follows the above pattern and there is a preponderance of the *syn* isomer ((**95**) to (**96**)). A model for predicting the outcome of these yeast reductions is shown in Scheme 27.

Scheme 26 Prelog's rule for the asymmetric microbial reduction of ketones.

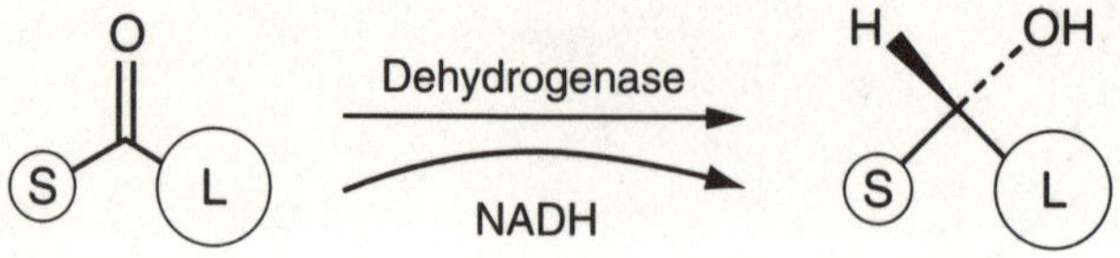

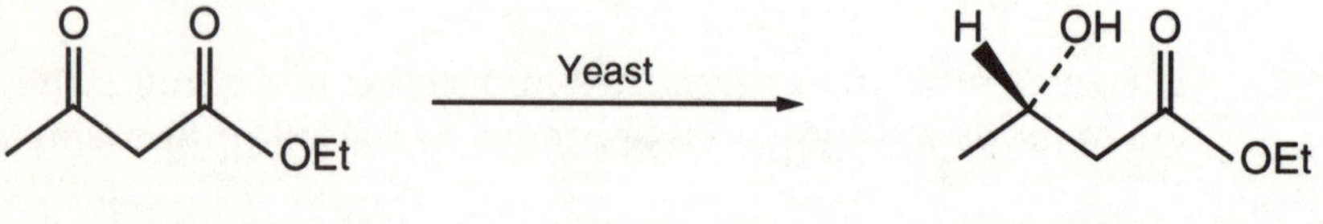

Dehydrogenase

NADH

91 → Yeast → 92

93 → Yeast → 94

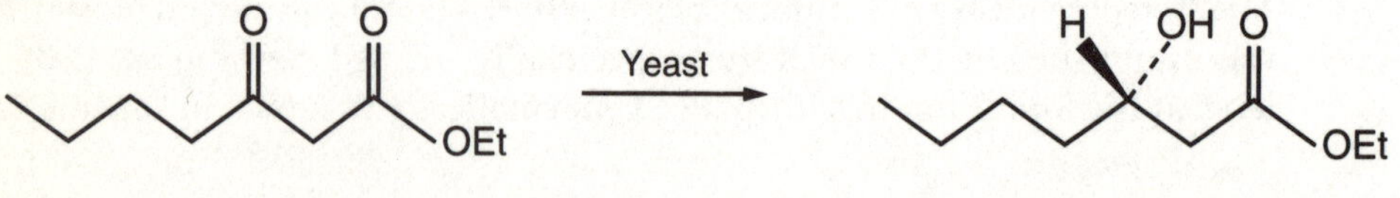

95 → Yeast → 96

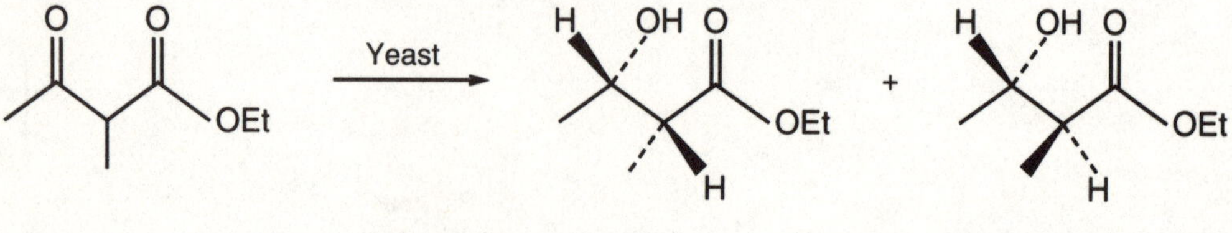

Scheme 27 A model for the influence of substituents on the reduction of ketones.

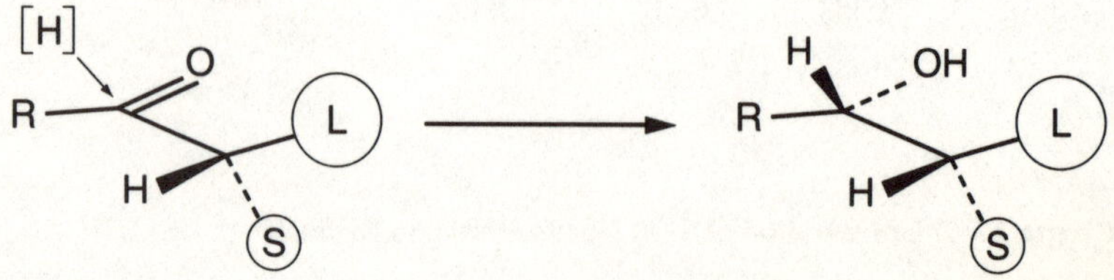

Reduction of cyclic ketones

The reduction of cyclic ketones takes place in a useful and predictable manner. Some examples are shown in Schemes 28 to 33.

Scheme 28

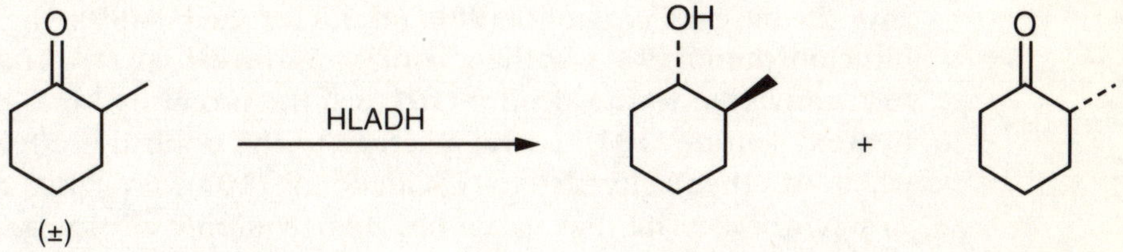
O
HLADH
(±)
OH
+
O

Scheme 29

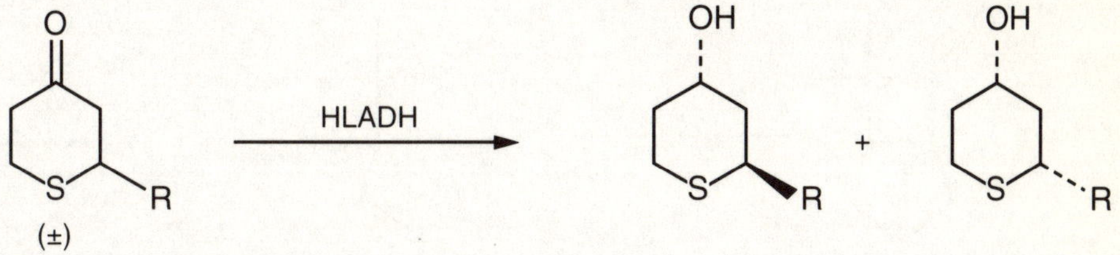
O
S
R
HLADH
(±)
OH
S
R
+
OH
S
R

Scheme 30

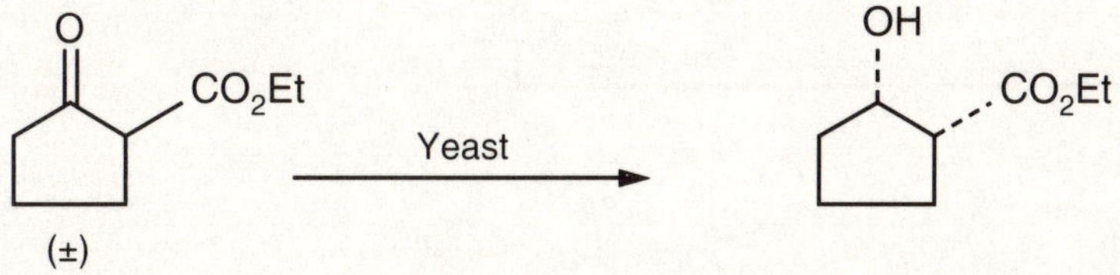
O
CO_2Et
Yeast
(±)
OH
CO_2Et

Scheme 31

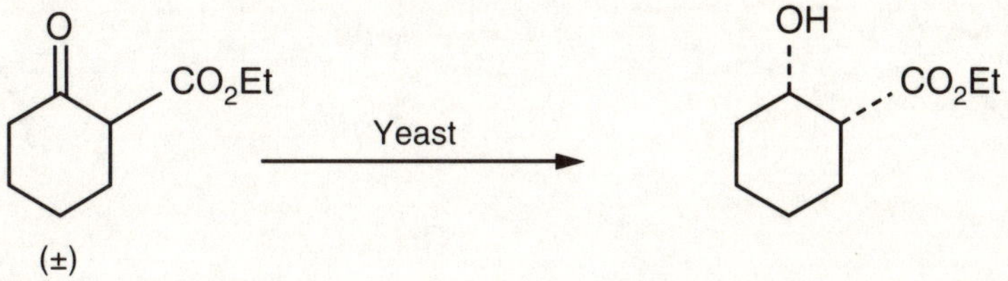
O
CO_2Et
Yeast
(±)
OH
CO_2Et

Scheme 32

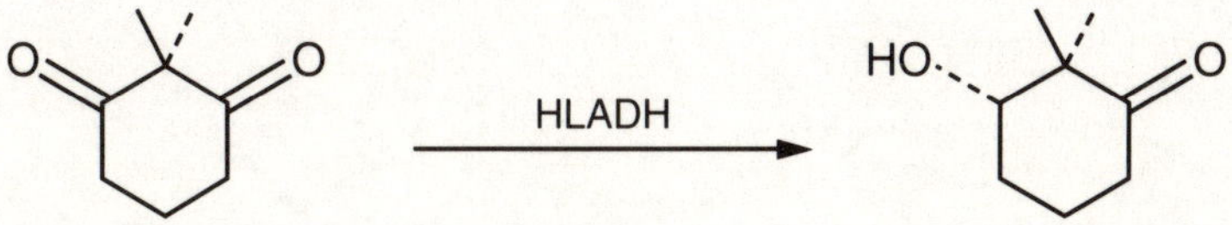
O
O
HLADH
HO
O

Scheme 33

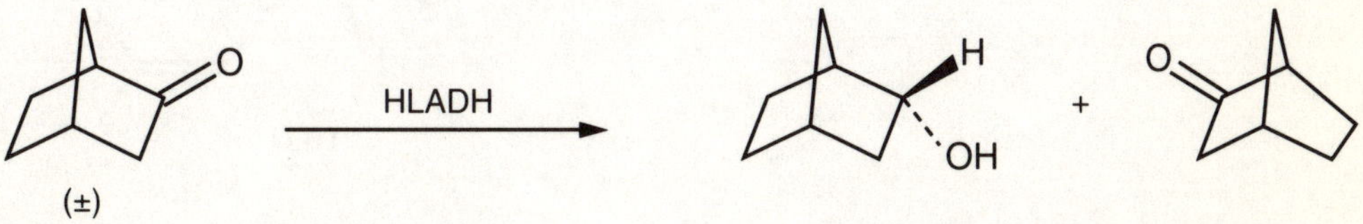
O
HLADH
(±)
H
OH
+
O

An application of this in synthesis is illustrated by the synthesis of the attractant pheromone, (+)-eldanolide (**103**). This is the attractant of a West African insect pest, the sugar cane borer, *Eldana saccharina*. The key stage involving the introduction of chirality was the reductive resolution of the ketone (**97**) using 3α,20-steroid dehydrogenase and NADH. The co-factor was recycled by coupling the reduction to the oxidation of ethanol by yeast alcohol dehydrogenase. This gave the pure 6–(*S*)–alcohol (**98**) and the ketone (-)–(**99**). The optically active bicycloheptenol (**98**) was then oxidized at the ketone (+)–(**100**), which was in turn converted to the alcohol (**101**) via the bromohydrin. Photolysis of this hydroxy-ketone (**101**) led via a ketene to the optically active lactone (**102**), which was then converted to (+)-eldanolide (**103**). Several points emerge from this synthesis. The first is the obvious economic advantage of introducing

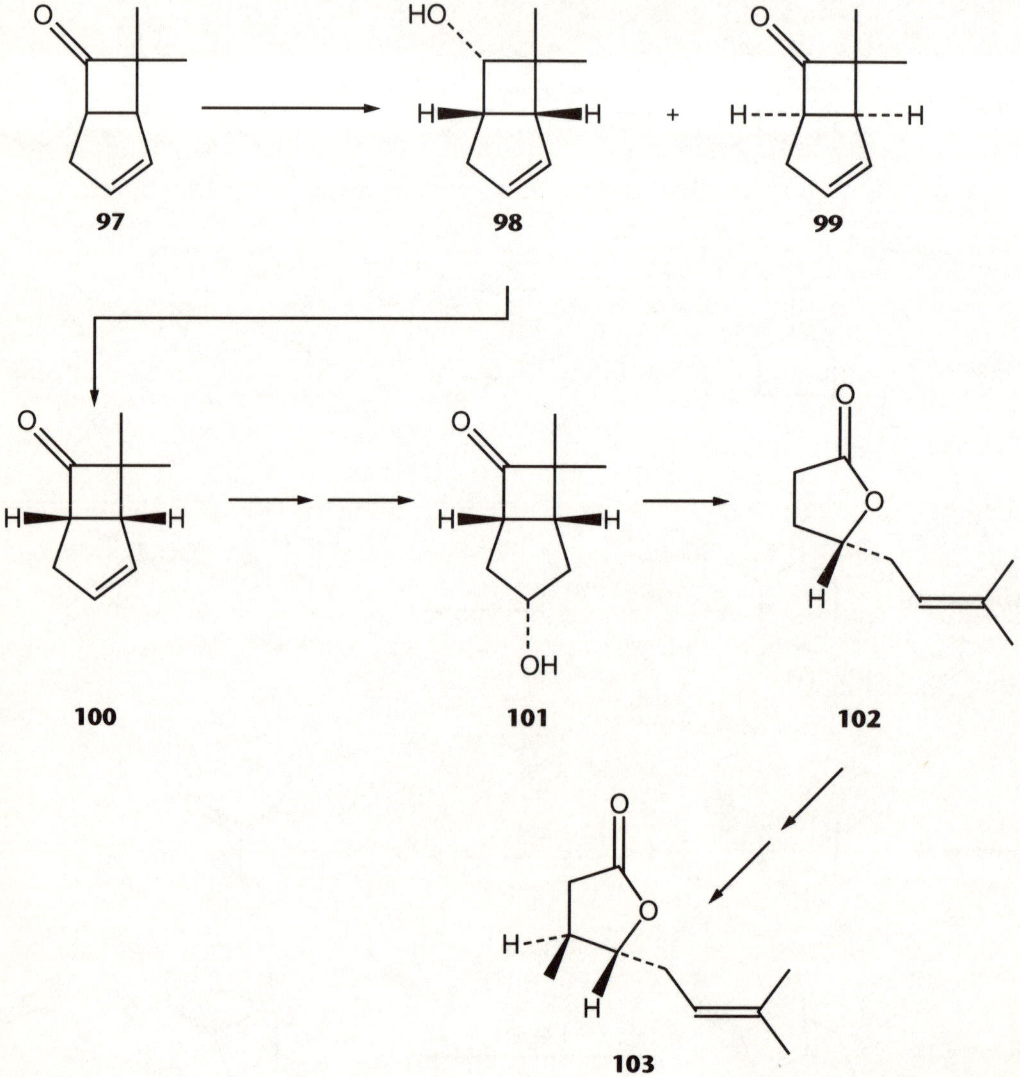

O HO O
H H + H H
97 98 99
O O O
H H H H O
H
OH
100 101 102
O
O
H
H
103

chirality at an early stage in the synthesis. The second is that the original chiral centre may not necessarily remain within the molecule but may be used to induce chirality elsewhere.

X-ray crystallographic studies have shown that HLADH is a dimer with a narrow barrel-like hydrophobic active site. The earlier diamond lattice model proposed by Prelog and the more recent cubic space model that has been proposed by J.B. Jones to accommodate the observed stereoselectivity, are compatible with this. The cubic lattice model is built up of 1.3 Å cubes (see Figure 4.1). The oxygen of the substrate alcohol or ketone is placed so that it can bind to a zinc atom, whilst the front of the cubic section is occupied by the 'A' face of the nicotinamide co-enzyme and in particular by C–4 of the nicotinamide, which delivers the hydrogen atom. This locates the CH(OH) group. The cubes are divided into those that can accept the substrate and those regions that are forbidden (i.e. regions that are already occupied by the aminoacids that constitute the structure of the enzyme). In the diagram the forbidden cubes are A1, A2, E1–3, G4, G5, L4–6, M7–8, Q7–9 and the left hand halves of B1, H4 and N7. Cubes O9, P8 and P9 are also limited. Those below 'forbidden' space are also forbidden, whilst those below limited or allowed space will accomodate some limited penetration by the substrate. Potential binding of the substrate must not violate any forbidden space (Figure 4.2). Comparability with the Prelog rule is shown in Figure 4.3.

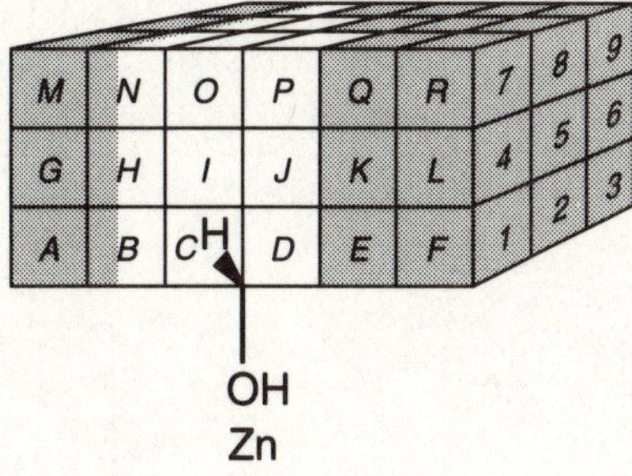

Fig. 4.1 A cubic lattice model for HLADH.

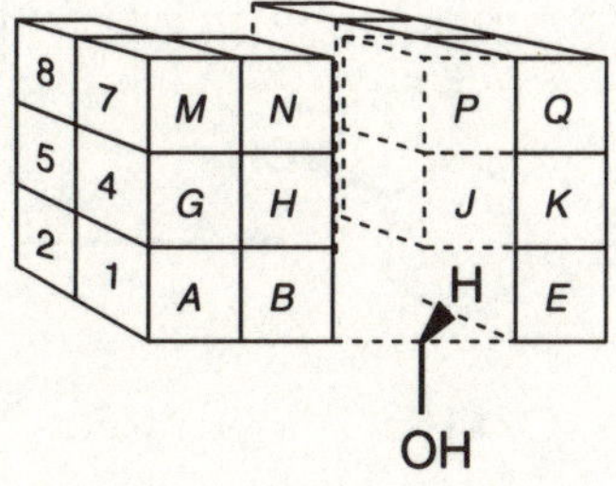

Fig. 4.2 'Allowed' spaces in the cubic lattice model of HLADH.

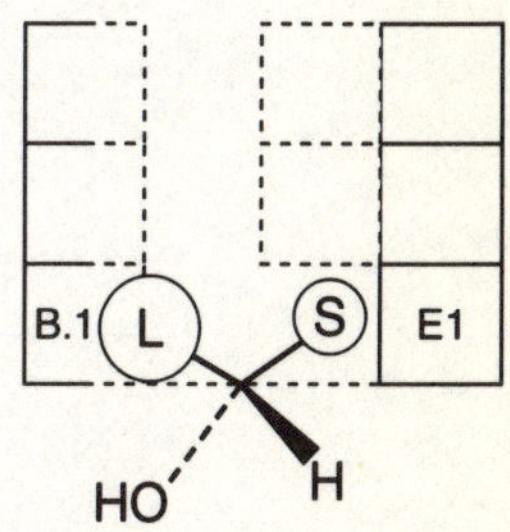

Fig. 4.3 A comparison between the Prelog model and the cubic lattice model of HLADH.

105

104

106

Unfavourable

Unfavourable

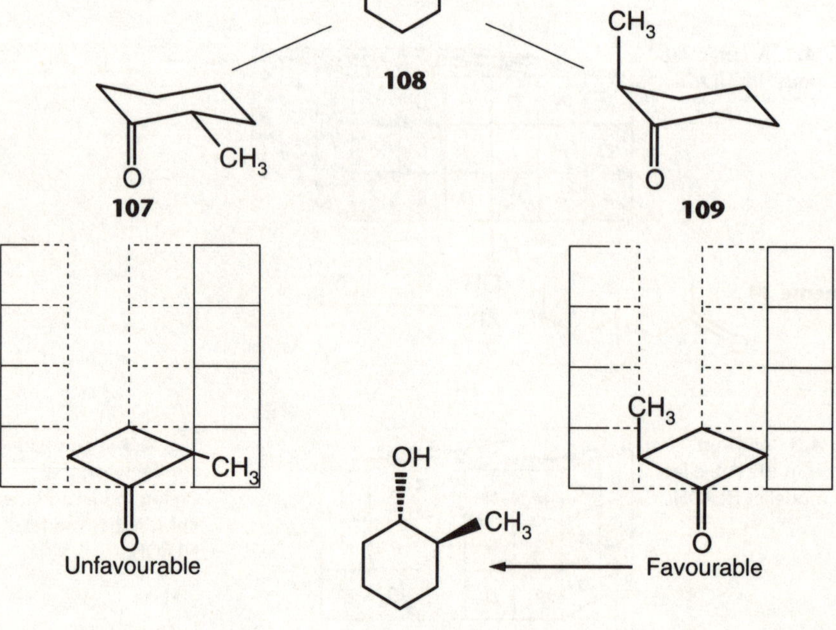

Fig. 4.4 The application of the cubic lattice model to the reduction of 2-alkylcyclohexanones.

The way in which the model operates may be seen in considering the reduction of the 2-alkylcyclohexanones (**105**) and (**108**) (Figure 4.4). The model predicts that reduction of the 2*R* series would be unlikely, since the binding of either of the conformers (**104**) and (**106**) would place the alkyl substituent in the forbidden regions of cubes B1 or K4. Reduction of the thermodynamically preferred conformer (**107**) of the 2*S* series with an equatorial alkyl group directed towards the forbidden E1 region is also unlikely. However the conformation (**109**) in the 2*S* series can bind since the 2-axial substituent can be accommodated within the allowed cube I4 and the limited H4 region. Reduction of this form of the bound substrate gives the *trans* (1*S*,2*S*)-2-alkylcyclohexanols (**110**). Experimentally, it is this isomer that is found and the 2*R*-alkylcyclohexanone is recovered unchanged.

The reduction of symmetrical bicyclic diketones may be effected selectively by HLADH. The carbonyl group that undergoes reduction does so with enantiotopic specificity. Thus the unsaturated decalin diones (**111**) and (**113**) are reduced to the corresponding hydroxy-ketones (**112**) and (**114**). This may be accompanied by chiral selection as exemplified by Schemes 34 and 35.

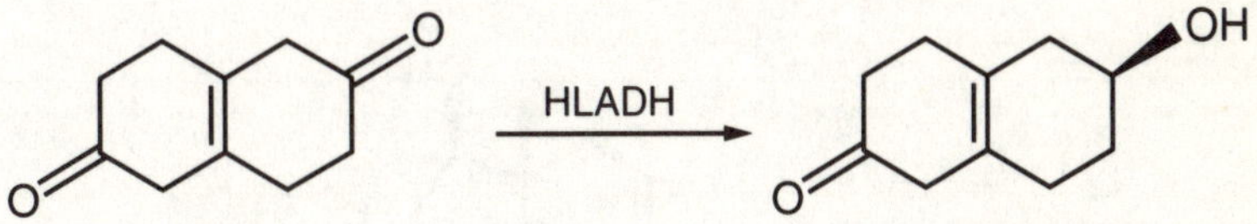

O O HLADH OH O

111 **112**

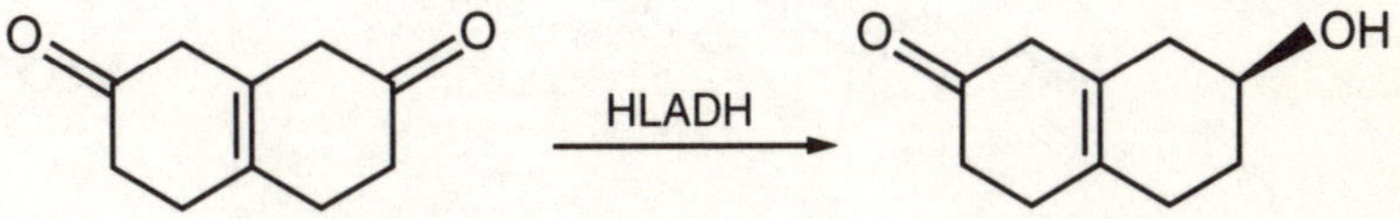

O O HLADH O OH

113 **114**

Scheme 34

H O H HLADH H OH O H + H O O H

(±)

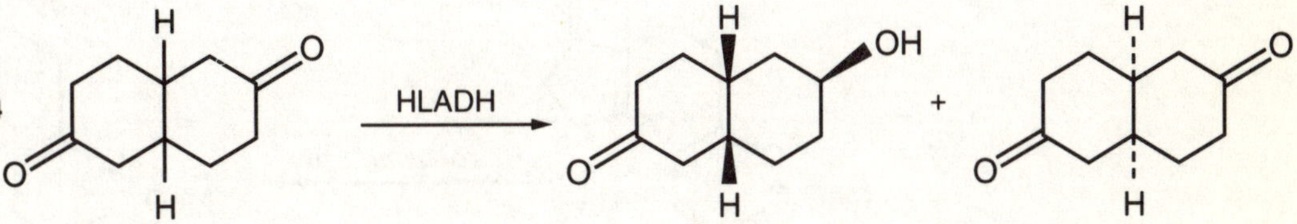

Scheme 35

H O H HLADH H OH O H

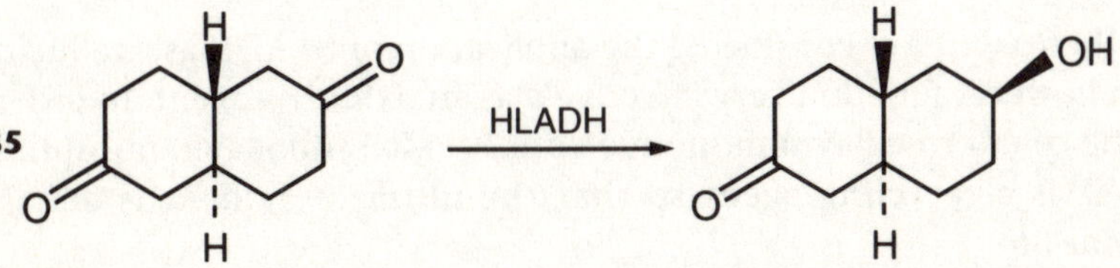

Different organisms may give different stereoisomers from the same substrate. This is exemplified by the reduction of the steroid precursor (**115**) by a yeast to afford (**116**) and by a bacterium to afford (**117**). The fungus *Rhizopus arrhizus* also selectively generates a steroid intermediate (**119**) in a useful chiral form from (**118**).

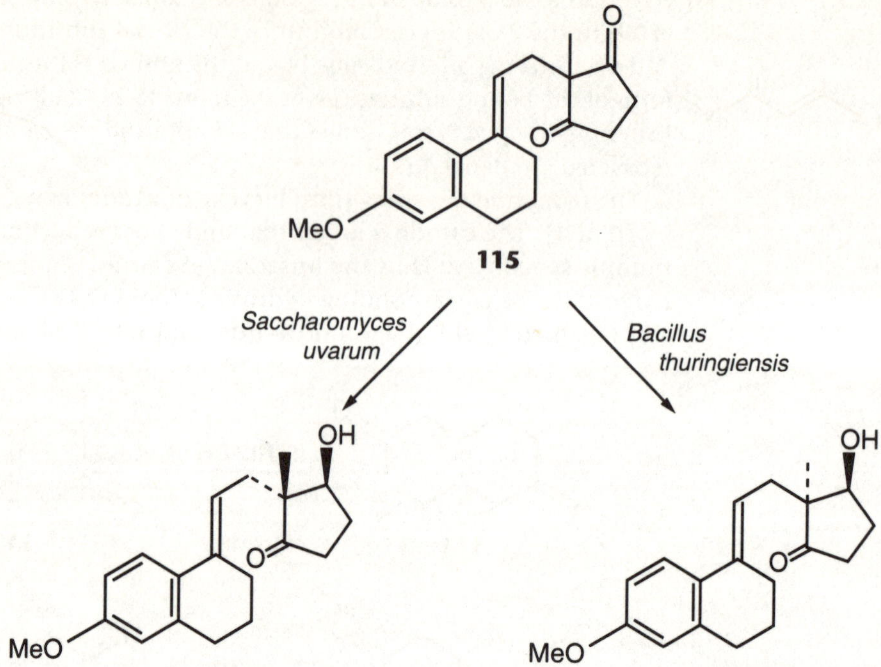

MeO
115
Saccharomyces uvarum
Bacillus thuringiensis
OH
MeO
116
OH
MeO
117

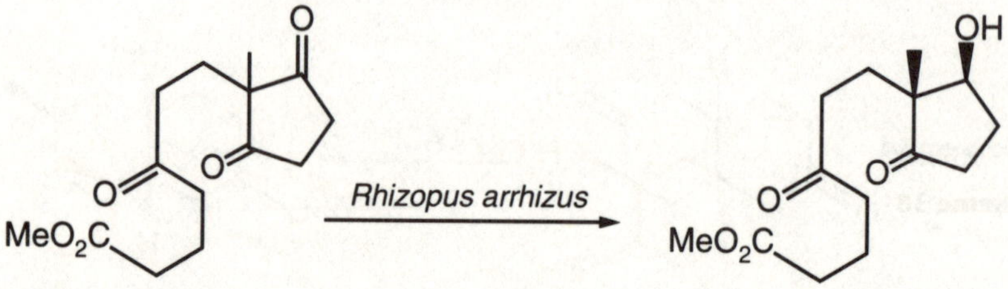

MeO$_2$C
118
Rhizopus arrhizus
OH
MeO$_2$C
119

Hitherto we have considered the application of HLADH as a reducing agent. It can however function very effectively as an oxidizing agent. In order to recycle, the co-enzyme flavin mononucleotide (FMN) (riboflavin phosphate) (**120**) is used as a recycling agent so that the ultimate oxidant is aerial oxygen (Scheme 36).

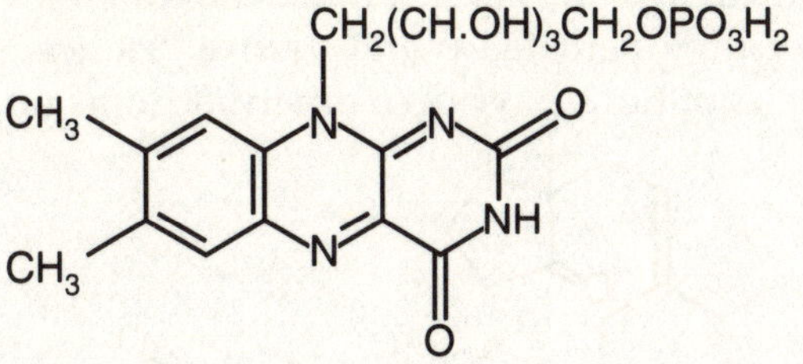

$CH_2(CH.OH)_3CH_2OPO_3H_2$

120 FMN

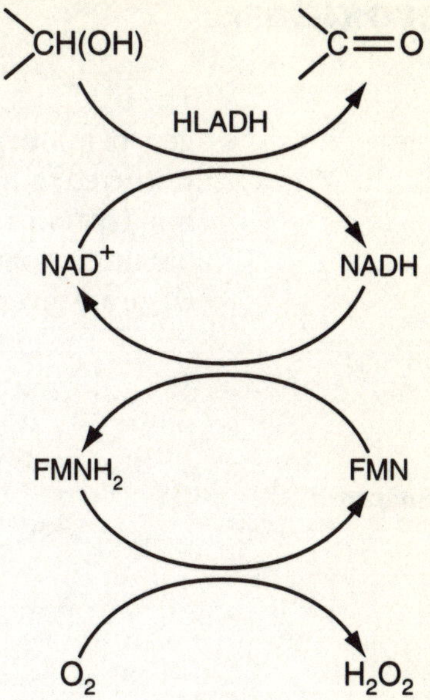

Scheme 36 Recycling the co-factor when HLADH is used as an oxidant.

A particularly revealing application of this involves the preparation of enantiomerically pure chiral lactones by the stereospecific HLADH oxidation of meso-diols. Here the ability of an enzyme to distinguish between constitutionally identical enantiotopic groups attached to centres of opposite chirality in a *meso* compound is utilized to good effect. Some examples are shown in the oxidation of the following diols (Schemes 37 to 39).

Scheme 37

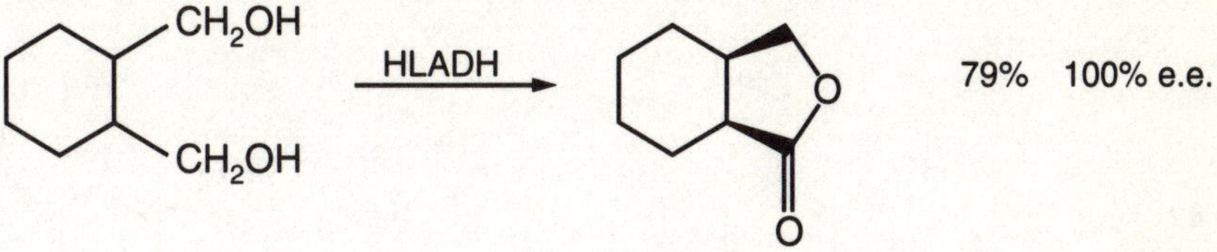

CH_2OH HLADH 79% 100% e.e.

Scheme 38

CH_2OH HLADH 80% 100% e.e.

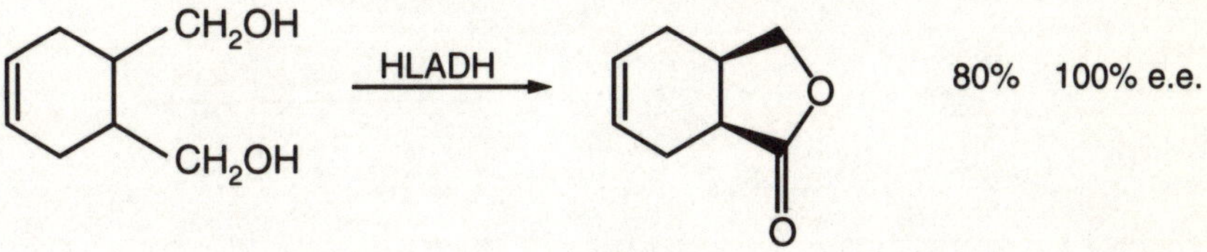

Scheme 39

CH_2OH HLADH 72% 100% e.e.

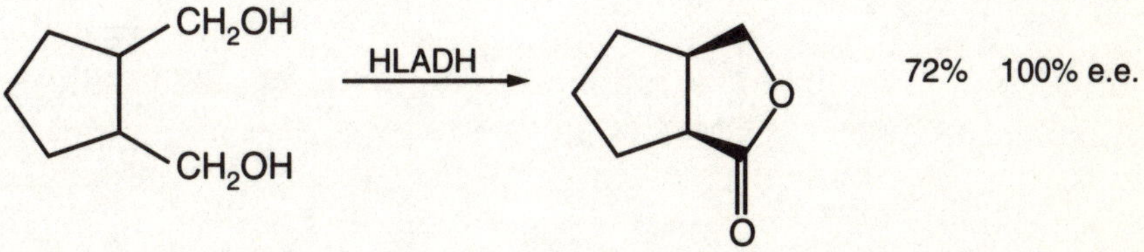

Peroxidase

The oxidation of phenols by peroxidase, particularly in the presence of hydrogen peroxide, has been known since 1903. Most of the early studies were restricted to the observation of colour changes. However, a number of interesting biotransformations with horseradish peroxidase and with a 'laccase' obtained from various *Polyporus* species of fungi have been observed. Some of these are given in Schemes 40 to 43.

Scheme 40

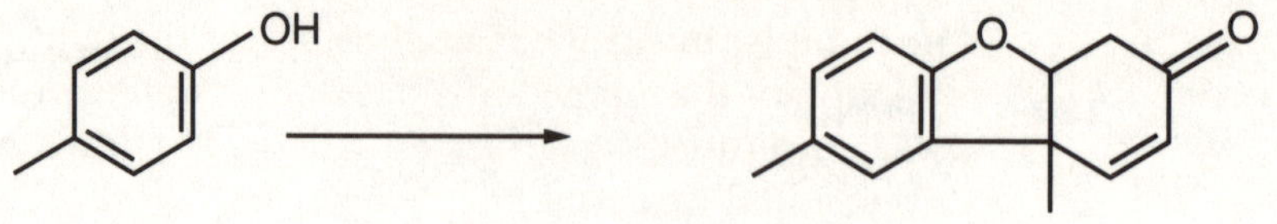

Scheme 41

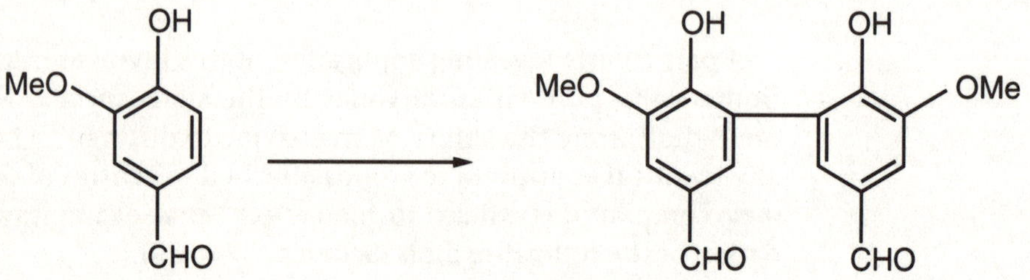

Scheme 42

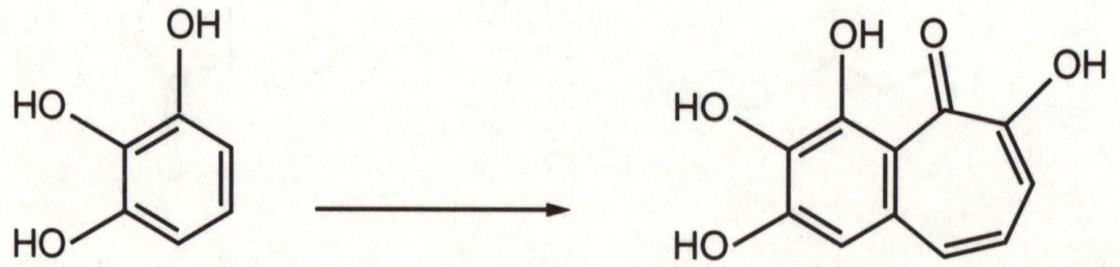

Scheme 43

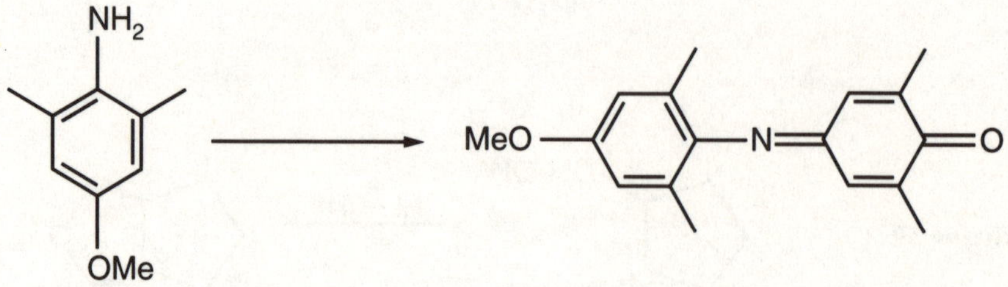

Most of the products arise by a one-electron transfer phenol coupling and similar oxidative processes exemplified by the well-known formation of Pummerer's ketone (see Scheme 44). Thus the dimerizations of phenols are typical radical processes.

Relatively few transformations have involved the insertion of oxygen. However, it has been found that horseradish peroxidase will catalyse the hydroxylation of some aromatic compounds by molecular oxygen in the presence of dihydroxyfumaric acid at 0ºC. This has been developed as a procedure for the preparation of L-DOPA (**122**) from L-tyrosine (**121**) and of 3,4-dihydroxyphenylglycine (**124**) from *p*-hydroxyphenylglycine (**123**).

Horseradish peroxidase contains a haem iron as in the cytochrome P_{450} system and it is this which is responsible for the radical formation. However, the formation of the many dimeric phenol coupling products, rather than those of oxygen insertion, suggests that there is a physical separation between the ferryl oxygen and the substrate derived radical.

Scheme 44 The formation of Pummerer's ketone.

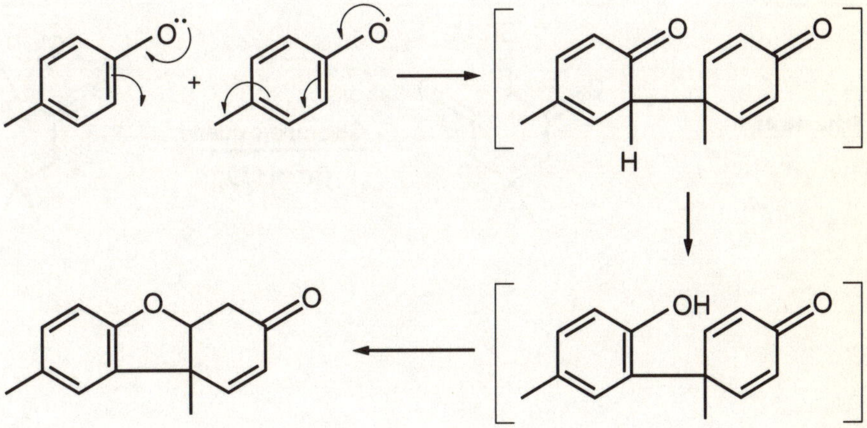

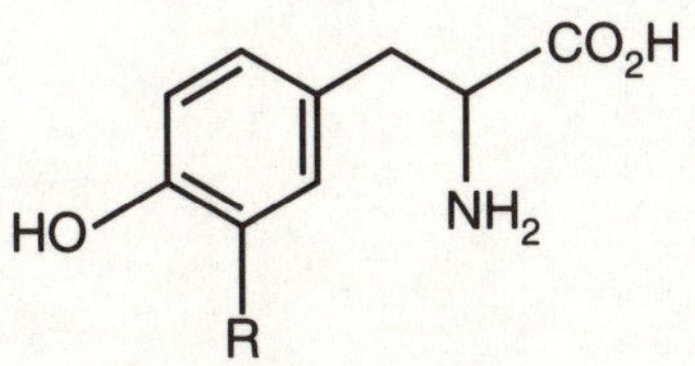

121 R = H
122 R = OH

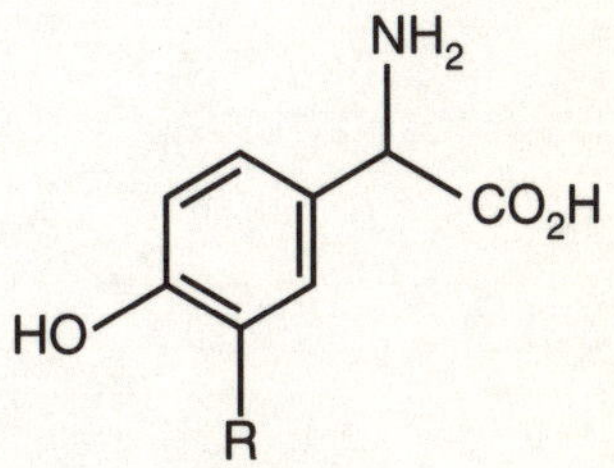

123 R = H
124 R = OH

Xanthine oxidase is another system that will bio-transform a range of substrate. It has been used in both an oxidative and a reductive sense, particularly with heterocyclic systems.

Chloroperoxidase is a similar system, which will halogenate electron-rich molecules, particularly enolic groups in the presence of hydrogen peroxide and an alkali metal halide. Typical bio-halogenations are exemplified by Schemes 45 and 46.

Scheme 45

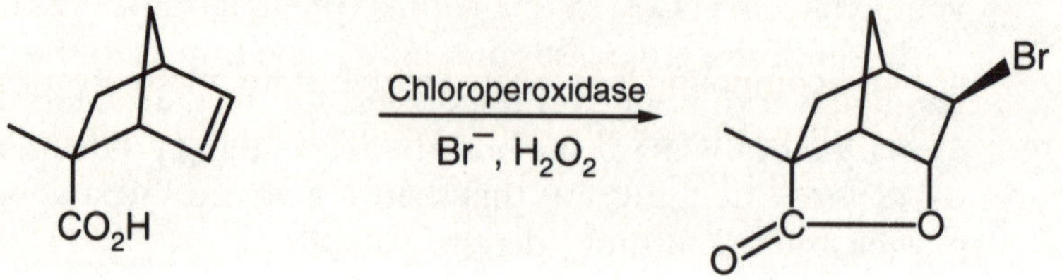

CO_2H

Chloroperoxidase

Br^-, H_2O_2

Br

O=C—O

Scheme 46

Cl

O O

Chloroperoxidase

Br^-, H_2O_2

Cl Br

O O

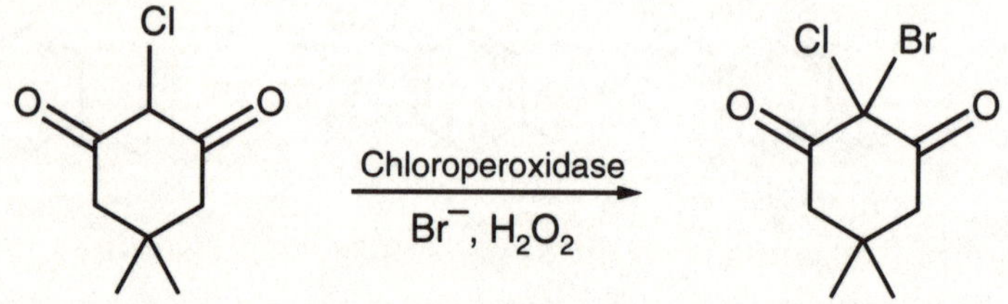

5 Microbiological hydroxylation and related reactions

When a compound is incubated with certain micro-organisms a hydroxyl group may be inserted at a site that is remote from other functionality. This ability of micro-organisms to hydroxylate compounds has a valuable synthetic potential. The intact organism is normally used for these transformations because many of the enzyme systems that are involved are membrane bound and are consequently difficult to obtain in a stable, isolated form. Those mono-oxygenase systems that have been studied appear not only to utilize cytochrome P_{450} but also require an electron source, which is usually NADPH. From a synthetic point of view, there is an advantage in terms of overcoming the need for co-factor regeneration by using whole organisms for these bio-transformations.

One enzyme system, however, which has been isolated and studied in considerable detail is the camphor hydroxylase of *Pseudomonas putida*. This catalyses the conversion of camphor (**125**) to 5-exo-5-hydroxycamphor (**126**). The prosthetic group for the enzyme is a cytochrome P_{450}, which is an iron–haem system (**127**). The sequence of events (Scheme 47) involves the reduction of the iron in the cytochrome P_{450} haem complex to iron(II), which takes up oxygen and a further electron and proton to form an iron (III)-OOH species. This loses a hydroxyl anion to form an iron(IV)-oxygen radical. In a two-step process this radical may abstract a hydrogen atom from the substrate to generate a carbon radical and an iron(IV)-hydroxy species. The carbon radical then accepts a hydroxyl radical from this to generate the hydroxylated product and iron(III).

The stereochemical outcome of these hydroxylations generally, but not exclusively, involves replacement of the hydrogen by a hydroxyl group with retention of configuration. An epoxide of the same stereochemistry is sometimes formed from an alkene at a site where the corresponding alkane is hydroxylated.

The hydroxylating cytochrome P_{450}-dependent enzymes are known as mono-oxygenases because they transfer one oxygen atom to the organic substrate. This is in contrast to the dioxygenases such as those utilizing α-keto-glutarate, in

O HO

125 **126**

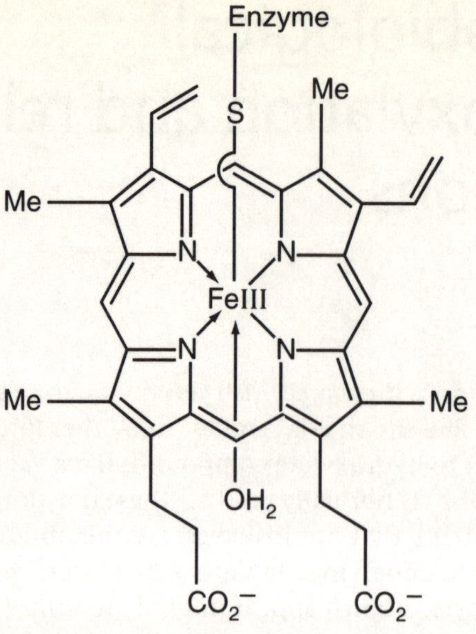

127

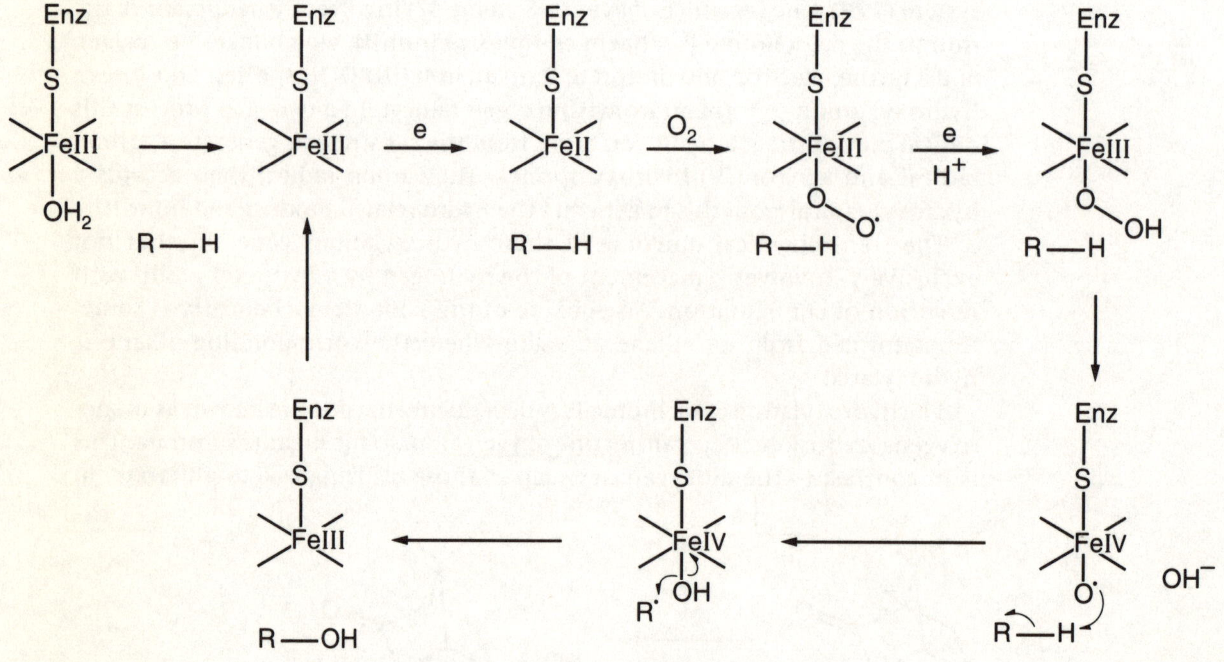

Scheme 47 Hydroxylation by a cytochrome P_{450}.

Cunninghamella elegans

Cephalosporium aphidicola

Beauveria sulfurescens

Ophiobolus herpotrichus

Rhizopus arrhizus

Scheme 48 Some microbiological hydroxylations.

which both oxygen atoms become attached to organic substrates. In their reactions with aromatic substrates, prokaryotic organisms such as bacteria utilize a dioxygenase to add oxygen to afford *cis*-diols whilst eukaryotic organisms such as fungi utilize mono-oxygenases, which afford arene oxides and benzylic oxidation products.

The value of microbiological hydroxylation lies in the fact that the sites of attack are often different from those at which chemical reactions occur and indeed some would be regarded as chemically remote. The examples in Scheme 48 illustrate this.

The microbiological hydroxylation of a very large number of substances has been examined using a variety of organisms and only a few examples can be discussed here. Incubation of both enantiomers of camphor (**132**, **133**) with the bacterium *Pseudomonas putida*, gave the 5-exo-alcohols (**129**, **130**) and the corresponding diketone (**128**) together with the lactone, campholide (**131**). These illustrate the typical pattern of microbial hydroxylation at a centre distant from a potential directing group and the microbiological equivalent of the Baeyer–Villiger reaction in the formation of the lactone. On the other hand, a Corynebacterium species has been found that will hydroxylate both enantiomers at the 6-position to give (**134**) and (**136**), from which the symmetrical diketone (**135**) was in turn obtained. The camphor hydroxylase of *Pseudomonas putida* has provided significant information on the structure of the enzyme.

O O **128** HO O **129** O OH **130**

O O **131** O **132** O **133**

HO O **134** O O **135** O OH **136**

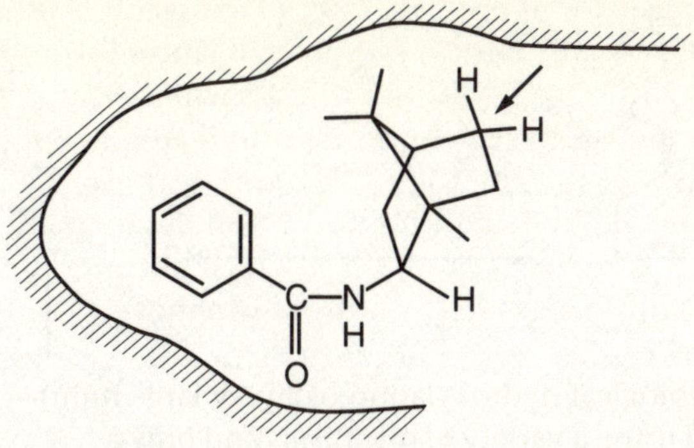

Fig. 5.1 A model for the active site of a hydroxylase in *Beauveria sulfurescens.*

Studies on the hydroxylation of some exo- and endo-bornylamides and their relatives by the fungus *Beauveria sulfurescens* at C–5 were consistent with a scheme in which there was a homolytic release of a hydroxyl radical from the cytochome P_{450}, which was able to abstract either the exo- or the endo-C(5)-hydrogen (see Figure 5.1), depending on the positioning of the substrate in the enzyme pocket. This positioning was determined by the anchoring role of the amide function. However, the second step involving the delivery of the iron-bound activated oxygen to the carbon radical at C–5 was stereospecific, taking place from an *exo*-direction. The model proposed by Furstoss has been used to rationalize the products of hydroxylation of a number of terpenoids.

The microbiological hydroxylation of steroids

Although the microbiological hydroxylation of steroids was known before 1952, it was the demonstration by Murray and Peterson that the fungus, *Rhizopus arrhizus*, would efficiently convert progesterone (**137**) into 11α-hydroxyprogesterone (**138**) that gave a major stimulus to this area. This biotransformation overcame a major problem in the synthesis of the cortical steroids, namely the insertion of an oxygen function at C–11, a position that is remote from other functional groups. This microbiological hydroxylation has formed the basis of a commercial process. Different micro-organisms will hydroxylate progesterone at other sites and in a number of cases will produce dihydroxylation (e.g. to form 6β, 11α-dihydroxyprogesterone (**140**)) and even cleave the side chain of progesterone to form testosterone acetate (**139**). The polyhydroxylations are time-dependent and thus *R. arrhizus* will afford mono- or di-hydroxylated products depending on the period of incubation. Some progress has been made in the purification of various progesterone hydroxylases from organisms such as *R. nigricans*, *Cochliobolus lunata*, *Aspergillus fumigatus* and *Mucor hiemalis*. Some of the more common steroid hydroxylations are shown in Figure 5.2.

The microbiological hydroxylation of steroids is not restricted to compounds related to progesterone. However, the site of hydroxylation may depend on the position of substituents. By systematically varying the position of carbonyl groups around the steroid framework, Sir Ewart Jones at Oxford was able to

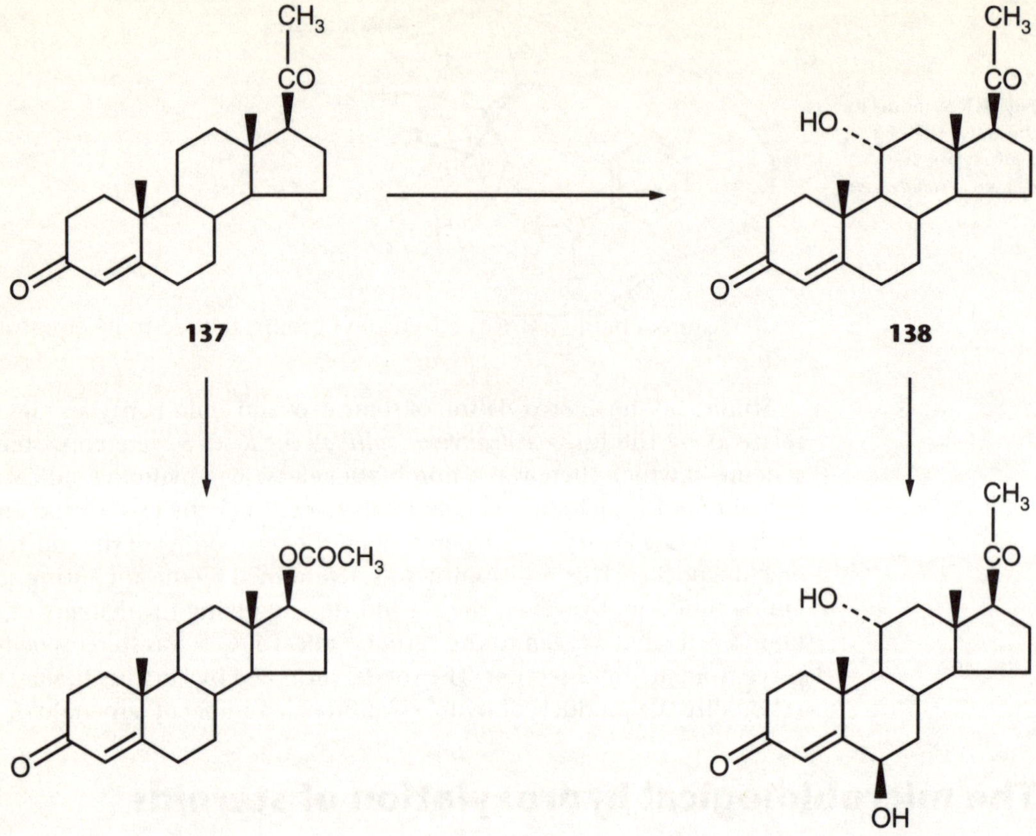

137

138

139

140

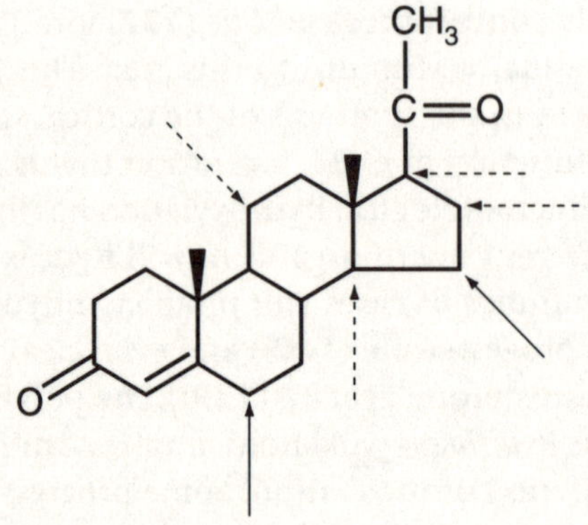

Fig. 5.2 The sites of the more common microbiological hydroxylations of progesterone.

establish some relationships between the site of hydroxylation and the position of a carbonyl group. For example, working with a series of monoketo-androstanes in which the carbonyl groups were separately placed at C–2, C–3, C–4 etc., the sites of dihydroxylation by the fungus *Calonectria decora* were shown to fall within an approximate geometrical relationship (see Figure 5.3).

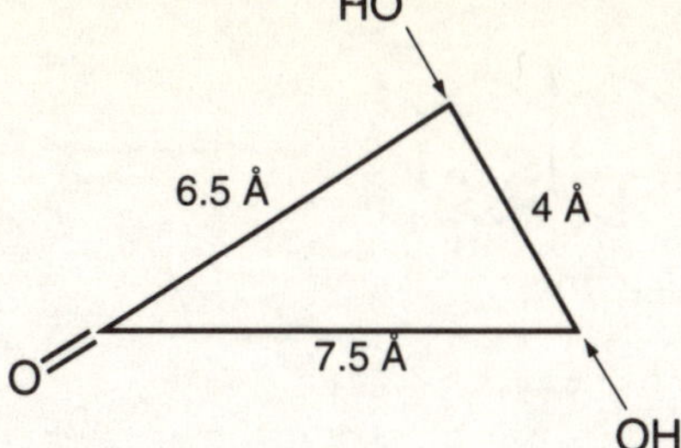

Fig. 5.3 Geometrical relationship between sites of microbiological dihydroxylation.

The stereochemistry of the hydroxyl groups tended to be equatorial. Some other organisms (e.g. *Rhizopus nigricans*) showed a pattern of hydroxylation that could also be accommodated in similar triangular relationships between a binding site and the sites of hydroxylation. Furthermore, the products from some mono-ketones could be accommodated by suggesting that they could bind in both a 'normal' (**141**) and/or a 'reverse' (**142**) manner. Thus a 3-ketone and a 16-ketone bear an approximate inverse relationship (see (**143**), Figure 5.4). This suggests that there may be some general framework for binding substrates to the hydroxylases.

There are some special situations. For example, a number of organisms such as *Rhizopus arrhizus* hydroxylate the Δ^4-3-keto steroids (**144**) at the axial 6β-position (**146**), and H.L. Holland has produced evidence that this proceeds through an enolic intermediate (**145**). Another feature that has been observed is the effect of an adjacent hydroxyl or halogen atom, which tend to inhibit microbiological hydroxylation at specific sites.

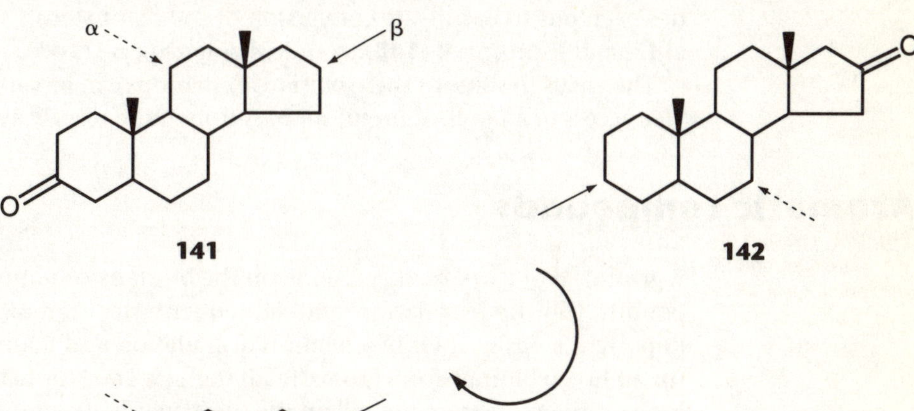

Fig. 5.4 The inverse relationship between a steroidal 3- and 16-ketone.

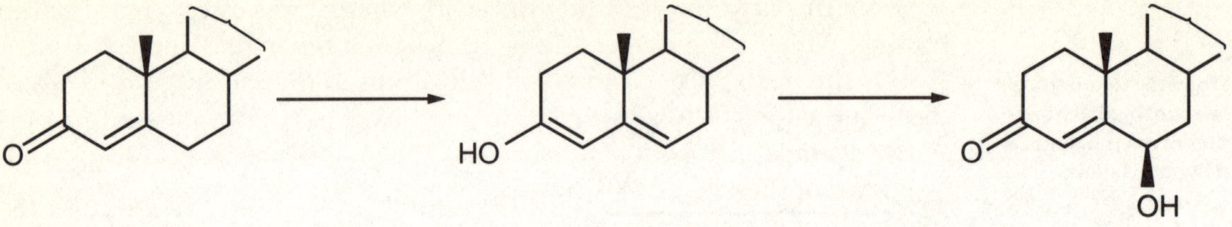

144 145 146

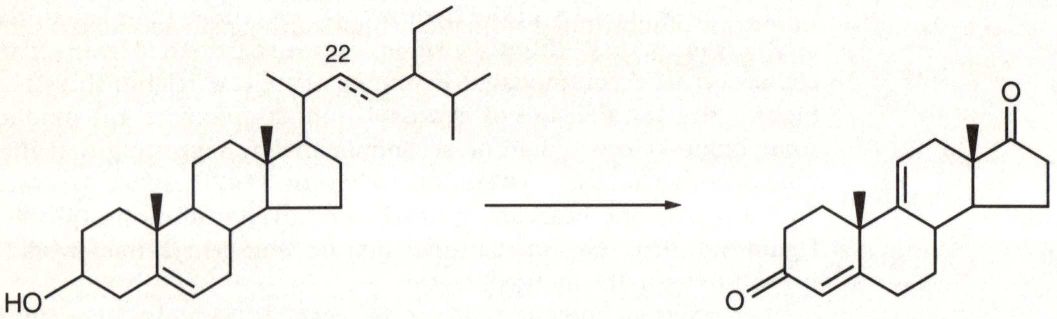

22

147 Δ^{22}

148 saturated

149

The bacterial hydroxylation and subsequent cleavage of the sterol side chain has been put to use in the conversion of the plant sterols such as stigmasterol (**147**) and β-sitosterol (**148**) to androstadienedione (**149**).

The yields in some of these biotransformations can be very high and they provide access to a range of useful steroids from more readily available materials.

Aromatic compounds

A great deal of work has been done on the biotransformation of aromatic compounds. The hydroxylation and subsequent cleavage of aromatic rings are important steps in their biochemical degradation and many organisms possess the ability to biotransform aromatic substances. Transformations may take place both on the aromatic ring and on the substituents. There are two distinct pathways for the oxidation of an aromatic ring. In the first, which is found with eukaryotic organisms (e.g. fungi), the addition of a single oxygen atom, takes place with the formation of an arene oxide (**151**). This reactive system may then undergo a number of reactions including hydrolysis to form a *trans*-1,2-diol (**150**), rearrangement (**153**) and subsequent enolization to generate a phenol (**154**) or reaction with another nucleophile (**152, 155**). The rearrangement pathway, which involves the shift of a hydrogen atom, is known as the NIH shift from the National Institutes of Health where it was discovered. The other pathway, which is characteristic of prokaryotic organisms such as bacteria, involves the introduction of two oxygen atoms to form a *cis*-1,2-dihydro-diol (**156**). Subsequent dehydrogenation may lead to an *ortho*-catechol (**157**). Further degradation involves cleavage to the *cis*-muconic acids (**160**) and the breakdown of these.

Mutants of the organism *Pseudomonas putida* have been found that lack the enzyme system that dehydrogenates the *cis*-1,2-diols (**156**) to the *ortho*-catechols. The *cis*-1,2-diols that then accumulate from substrates such as benzene (**161**), toluene (**162**) and chlorobenzene (**163**) have proved to be valuable and, in the case of substituted aromatics, chiral starting materials for the synthesis of a range of compounds. Some of these products are set out in Scheme 49. The particular enantiomer of the *cis*-dihydrodiol that is formed preferentially by *Pseudomonas putida* in the dioxygenase catalysed oxidations of 1,4-disubstituted benzenes is largely controlled by the relative size of the substituents, as shown in Scheme 50.

R

R R R

OH

OH O Nu

HO

150 **151** **152**

R R R

Nu

O OH OH

153 **154** **155**

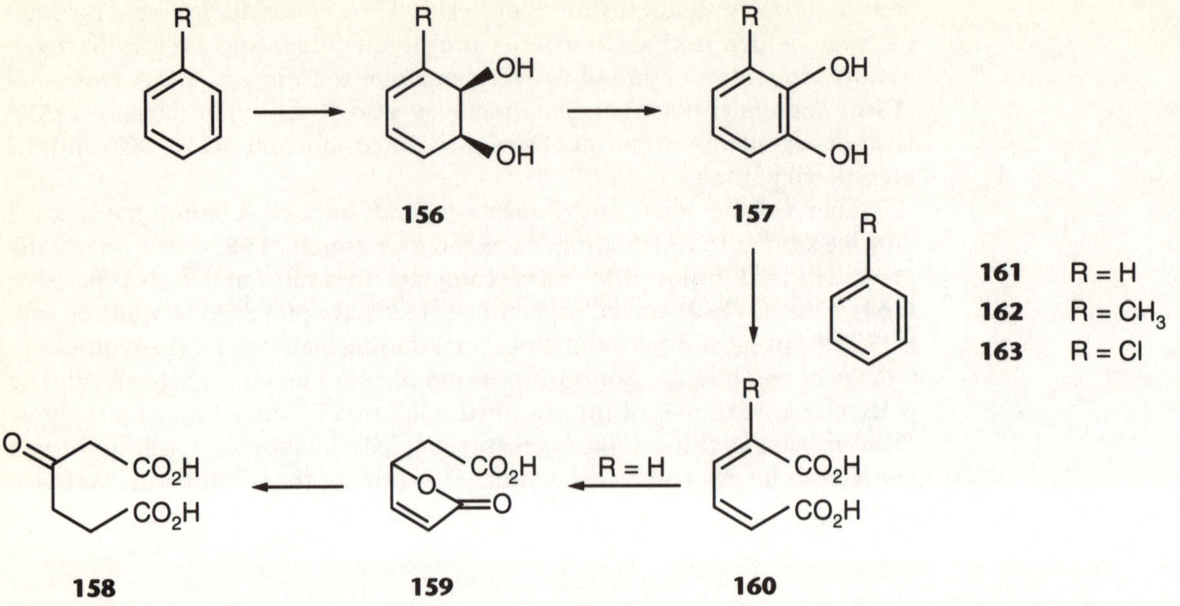
R
R
OH
OH
156
R
OH
OH
157
R
161 R = H
162 R = CH3
163 R = Cl
R
CO2H
CO2H
160
R = H
CO2H
O
O
159
O
CO2H
CO2H
158

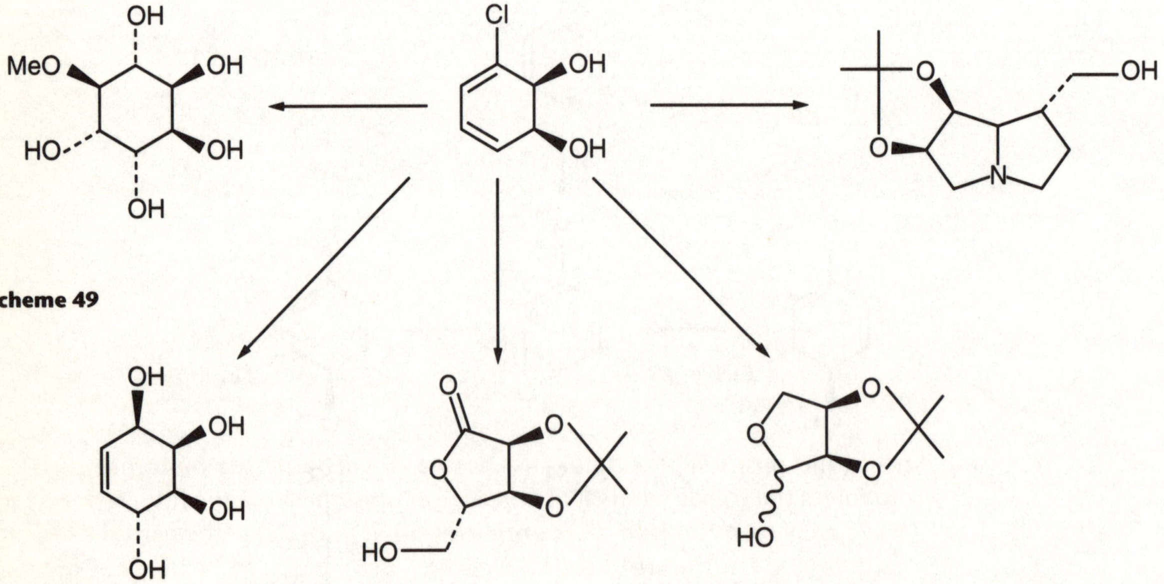
Cl
OH
OH
OH
MeO
OH
HO
OH
OH
O
O
OH
N
OH
OH
OH
OH
O
O
O
O
HO
O
O
O
HO

Scheme 49

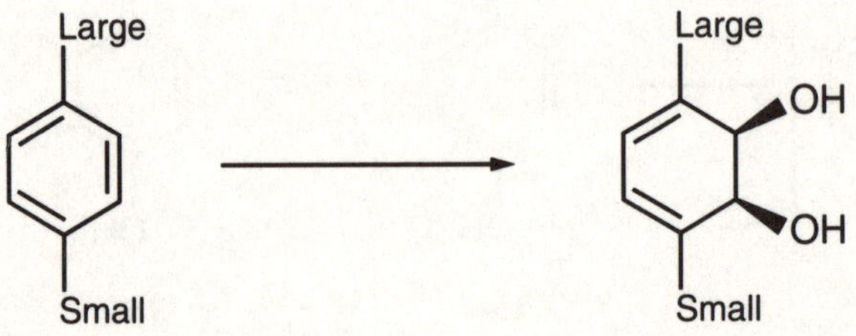
Large
Small
Large
OH
OH
Small

Scheme 50

The degradation of the catechols may follow one of a number of pathways leading, for example, via *cis*, *cis*-muconic acid (**160**) to muconolactone (**159**) and thence to 3-ketoadipate (**158**). The biotransformation of aromatic compounds may provide access to some of the substituted versions of these aliphatic compounds.

Oxidation at the benzylic position is more common with fungi. For example, whereas the rings of 2-methylnaphthalene (**164**) are degraded by bacteria, for example to (**165**), the methyl group is oxidized to naphthoic acid (**166**) by the fungus *Cunninghamella elegans*. The alkylnaphthalenes are found in the toxic residues of waste oil and their metabolites have been studied in this connection.

OH
HO
CH_3
HO
CH_3
HO
165
Pseudomonas desmolyticum
CH_3
164
Cunninghamella elegans
CH_2OH
CO_2H
166

The fungus *Mortierella isabellina* will hydroxylate aromatic compounds in the benzylic position. H.L. Holland has proposed that this biotransformation involves the initial removal of an electron from the aromatic ring, followed by the loss of a proton from the benzylic position to generate the benzylic radical. The product is then formed by attack of the oxygen from the iron–oxygen species of the cytochrome. A model has been proposed to accommodate the substrate selectivity and stereochemistry shown by the organism.

The enzymatic Baeyer–Villiger reaction

The result of an apparent enzymatic Baeyer–Villiger reaction is often observed in the biodegradation of a natural product. The opening of ring D of the steroids such as androstenedione (**167**) to afford testolactone (**168**) has been observed, accompanying various microbiological hydroxylations. However, there are a number of specific applications of the enzymatic Baeyer–Villiger reaction, as it offers the potential for the enantioselective desymmetrization of cyclic ketones. A cyclohexanone oxidase has been obtained from an Acinetobacter species and this has been applied to the desymmetrization of 4-substituted cyclohexanones, (**169**) to (**170**). With the isolated enzyme system, there is a requirement for co-factor (NADPH) regeneration. When the side chain is a hydroxymethyl group, the seven-membered lactone rearranges to afford a chiral δ-lactone (**171**). Other studies have been made using the (+)- and (–)-camphor mono-oxygenases induced in *Pseudomonas putida* in an effort to establish systems with contrasting enantioselectivity.

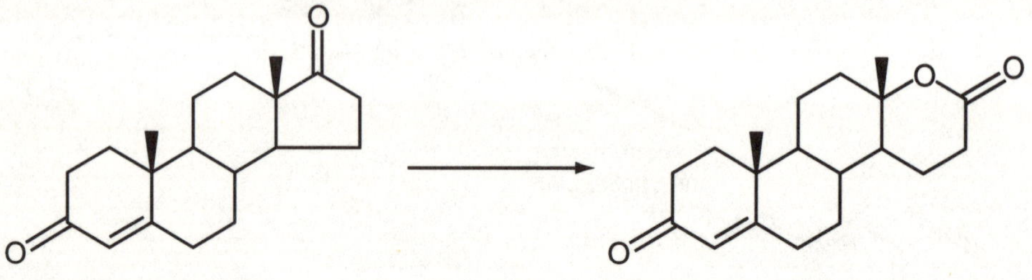

167 **168**

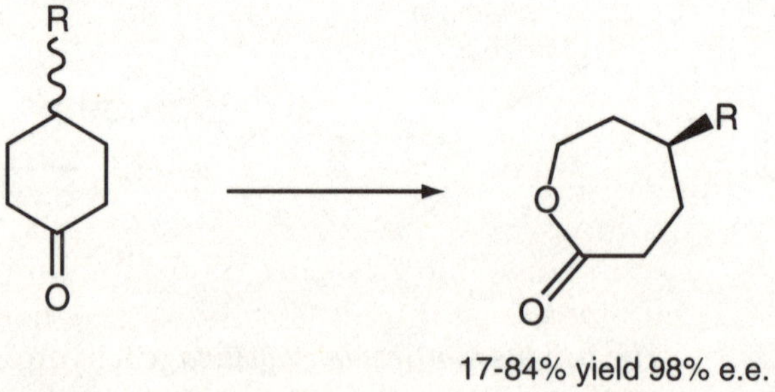

R

O

169

R

O

O

17-84% yield 98% e.e.

170

CH_2CH_2OH

O

O

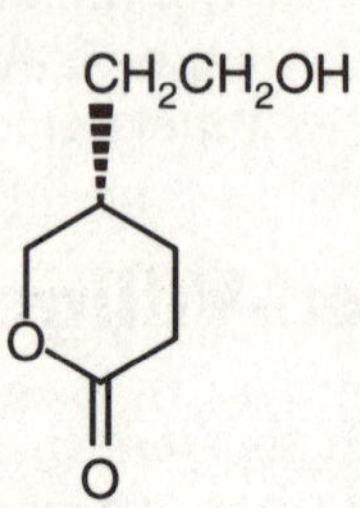

171

6 Biosynthetically-directed biotransformations

Biosynthetically-directed biotransformations have played an important role in a number of studies. On the one hand, the substrate flexibility of specific enzyme systems can be used to reveal aspects of the enzyme mechanism, whilst on the other hand it can be used to produce analogues of biologically active compounds. Whereas it is part of the function of many of the catabolic enzymes used in xenobiotic transformations to have a wide substrate specificity, this is not necessarily the case for enzymes on biosynthetic pathways. Consequently, the efficiency and overall yield of biosynthetically-directed transformations are often disappointing. It is the information these transformations can provide that is their redeeming feature. In this chapter we will consider some examples of the application of biosynthetically-directed biotransformations in terms of the strategies used to obtain biosynthetic information, the production of novel metabolites and the design of selective biosynthetic inhibitors.

2,3-Epoxysqualene–lanosterol cyclase

A proposal for the course of the cyclization of squalene and the rearrangements that lead to lanosterol was made by Woodward and Bloch in 1953, whilst the intermediacy of 2,3-epoxysqualene (**172**) was first demonstrated by Corey and by van Tamalen in 1966. It was originally proposed that the cyclization of (**172**) gave a protosterol carbonium ion with the stereochemistry (**173**). This has subsequently been modified to (**174**). The cyclization was followed by a series of rearrangements that were completed by the loss of the 9β-hydrogen atom and the formation of lanosterol (**175**) (see Scheme 51). Incubation of a series of analogues of 2,3-epoxysqualene with enzyme systems derived from yeast and hog liver have provided information on the underlying structural requirements of the cyclase. The analogues comprised variants of 2,3-epoxysqualene lacking double bonds or methyl groups, partially-cyclized versions of 2,3-epoxysqualene and compounds containing auxiliary functions that might trap a carbocation. The terminal isoprene unit of the epoxide was not essential for the cyclization since the analogue (**176**) was converted efficiently to the tetracyclic product (**177**). 18,19-Dihydro-2,3-epoxysqualene (**178**) was accepted by the cyclase but converted to a tricyclic product (**179**). Use of desmethyl analogues revealed that the methyl groups adjacent to the epoxide were important,

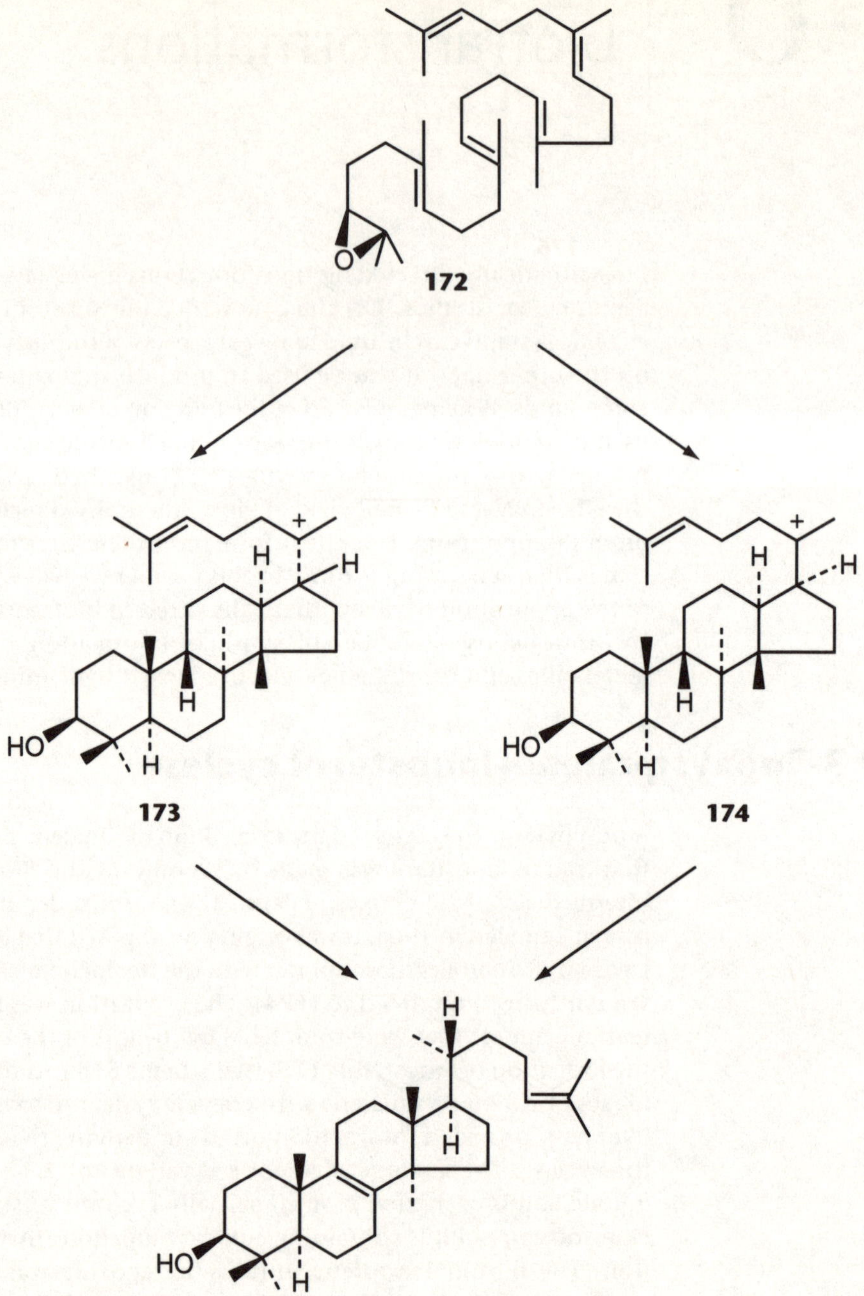

172

173

174

175

Scheme 51 The enzymatic cyclization of 2,3-epoxysqualene.

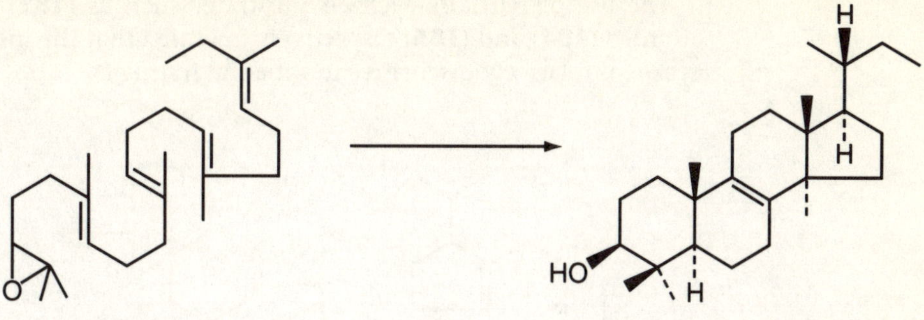

176 177

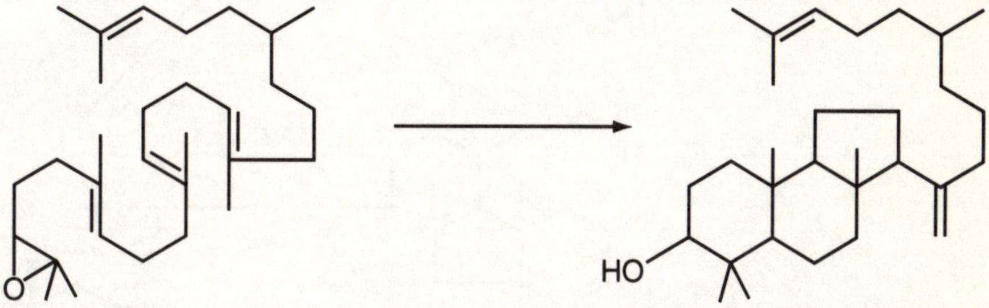

178 179

possibly for orientation of the substrate by the cyclase, whilst the central methyl groups at C-6, C-10 and C-15 of 2,3-epoxysqualene on their own were less important. However, when both the C-10 and C-15 methyl groups were missing (**180**), the product was (**181**) indicative of a role for these methyl groups possibly through their interaction with the enzyme system to produce the correct folding of 2,3-epoxysqualene.

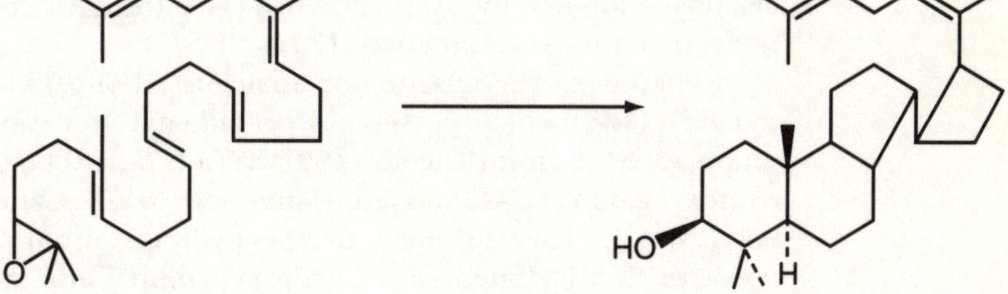

180 181

The use of partially-cyclized substrates such as (**182**) and (**183**), which afforded (**184**) and (**185**) respectively, revealed that the initial reaction of the cyclase is relatively insensitive to variation in ring D.

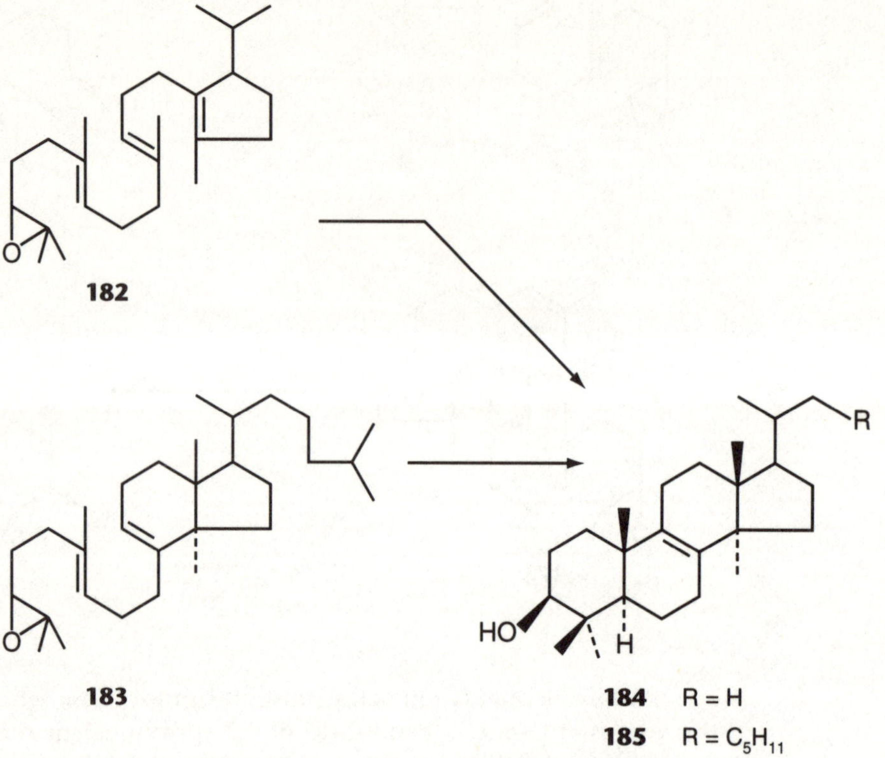

182

183

184 R = H

185 R = C_5H_{11}

Another structural variation that proved rewarding was to build into the 2,3-epoxysqualene cation stabilizing auxiliaries to trap specific intermediates. Incubation of the 20-oxa analogues of 2,3-epoxysqualene (**186**) afforded the 17β-acetyl compounds (**187**). Tritium-labelling showed that the hydrogen atom at C–17 had remained in place throughout the cyclization. This provided a clear demonstration that the cyclization had taken the steric course leading to (**174**) rather than the earlier proposal (**173**).

Incubation of the unsaturated analogue (20*E*)–20,21-dehydro-2,3-epoxysqualene (**188**) which possesses a pentadienyl unit capable of stabilizing a cation, gave the protostanediol (**189**) with a 17β-side chain. The addition of the hydroxyl group at C–20 was fast relative both to any rotation of the C–17-C–20 bond and to rearrangement of the hydrogen from C–17 to C–20. The regiospecific attachment of the hydroxyl group is also interesting in that it is possible that in the natural cyclization it is the enzyme-bound water that stabilizes the protosterol C–20 cation formed from 2,3-epoxysqualene prior to the rearrangement steps.

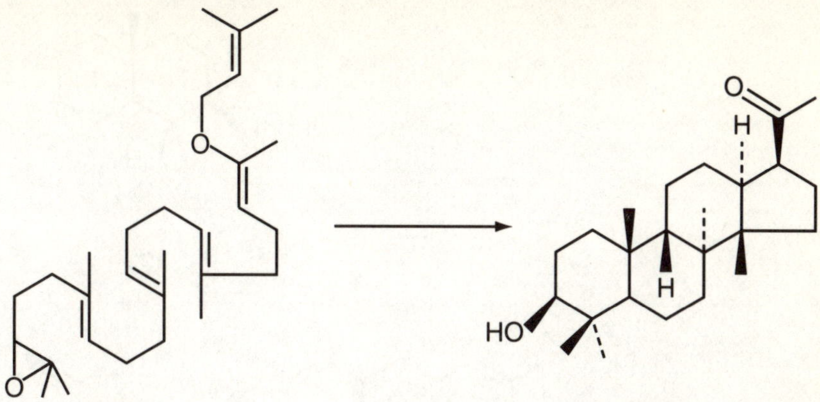

O
H
H
HO
186
187

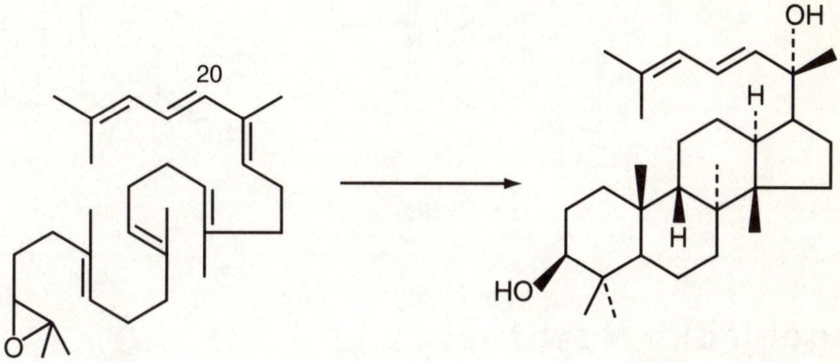

OH
20
H
H
HO
188
189

These analogue biosynthetic transformations led to the development of a series of 2,3-epoxysqualene cyclase inhibitors. These include substrate and transition state mimics such as 2,3-iminosqualene (**190**), the aza analogues (**191**) and their related *N*-oxides, product mimics such as the decalol (**192**) and the mechanism-based 29-methylidene-2,3-epoxysqualene (**193**) inhibitor. The latter was possibly irreversibly bound to the cyclase by an active site nucleophile, which was able to attack the allylic carbocation (**194**).

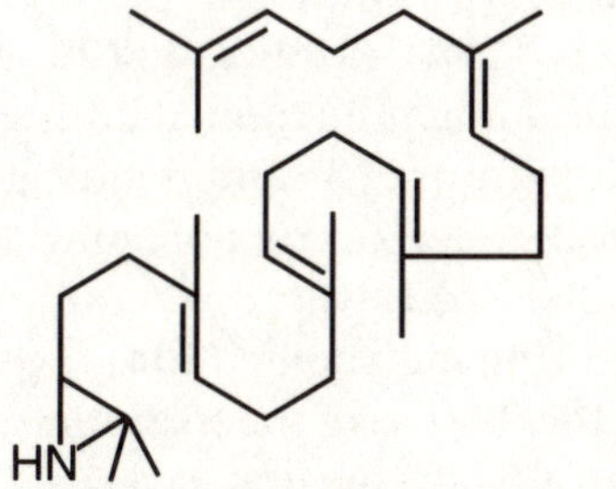

HN
N
190
191

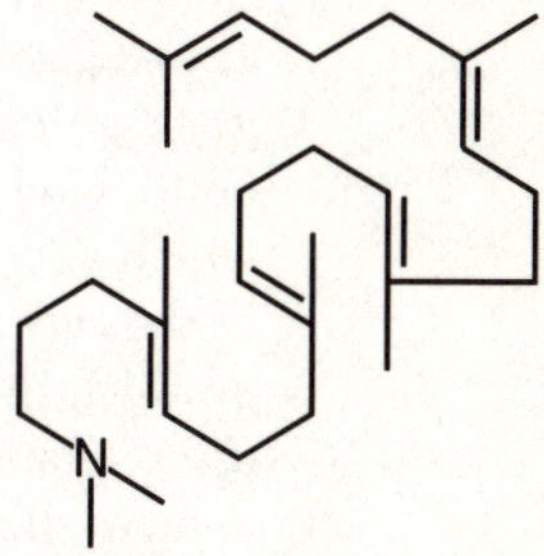

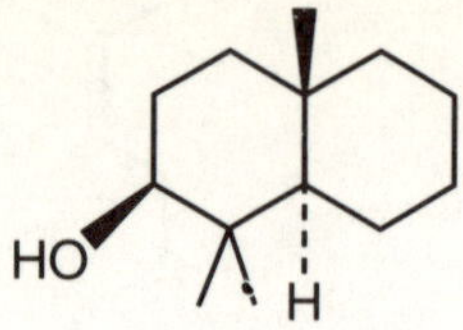

HO H

192

Enzyme Enzyme H+ O H HO

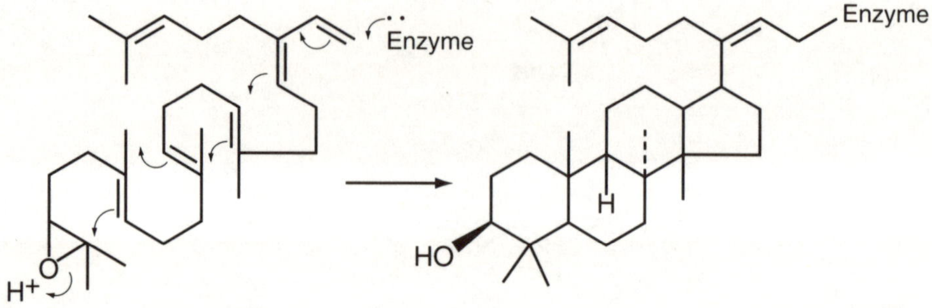

193 194

Isopenicillin N synthase

Isopenicillin N synthase is the key enzyme system that is responsible for the conversion of the tripeptide, δ-(L-α-aminoadipoyl)-L-cysteinyl-D-valine (LLD-ACV) (**195**) into isopenicillin N (**196**) and thence the penicillin antibiotics. The enzyme system has been purified from and cloned into *E. coli*. A systematic programme by Baldwin, involving modification of the components of the tripeptide, has been particularly rewarding in mapping the active site and providing insight into the mechanism of formation of the penicillin ring system. Amongst these studies, replacement of the valine component utilized some general strategies and gave insight into biosynthetic reactions with a radical character.

Replacement of the valine with aminobutyrate (**197**) labelled with deuterium at C–3 gave three products, (**198**), (**199**) and (**200**), retaining deuterium, suggesting that a carbon radical might be formed during the formation of the C–S bond. One test for the formation of a carbon radical is based on the ease with which a cyclopropyl radical undergoes ring opening, (**201**) to (**202**). Incubation of the cyclopropyl derivative (**203**) with the enzyme system gave, as the major product, the 8-membered ring metabolite (**204**) together with minor products (**205**) retaining the ring. Incubation of the tripeptide (**206**) obtained by replacement of the valine with an allylglycine unit gave five products (**207** to **211**), two arising by an oxygenation process and three arising from dehydrogenation. The formation of the oxygenation products may be accounted for by interaction

195

196

198

199

197

200

AA =

with a ferryl (IV)=0 species. These results, which exemplify the strategy of analogue biosynthesis, taken with a considerable body of other evidence, lead to the catalytic cycle given in Scheme 52.

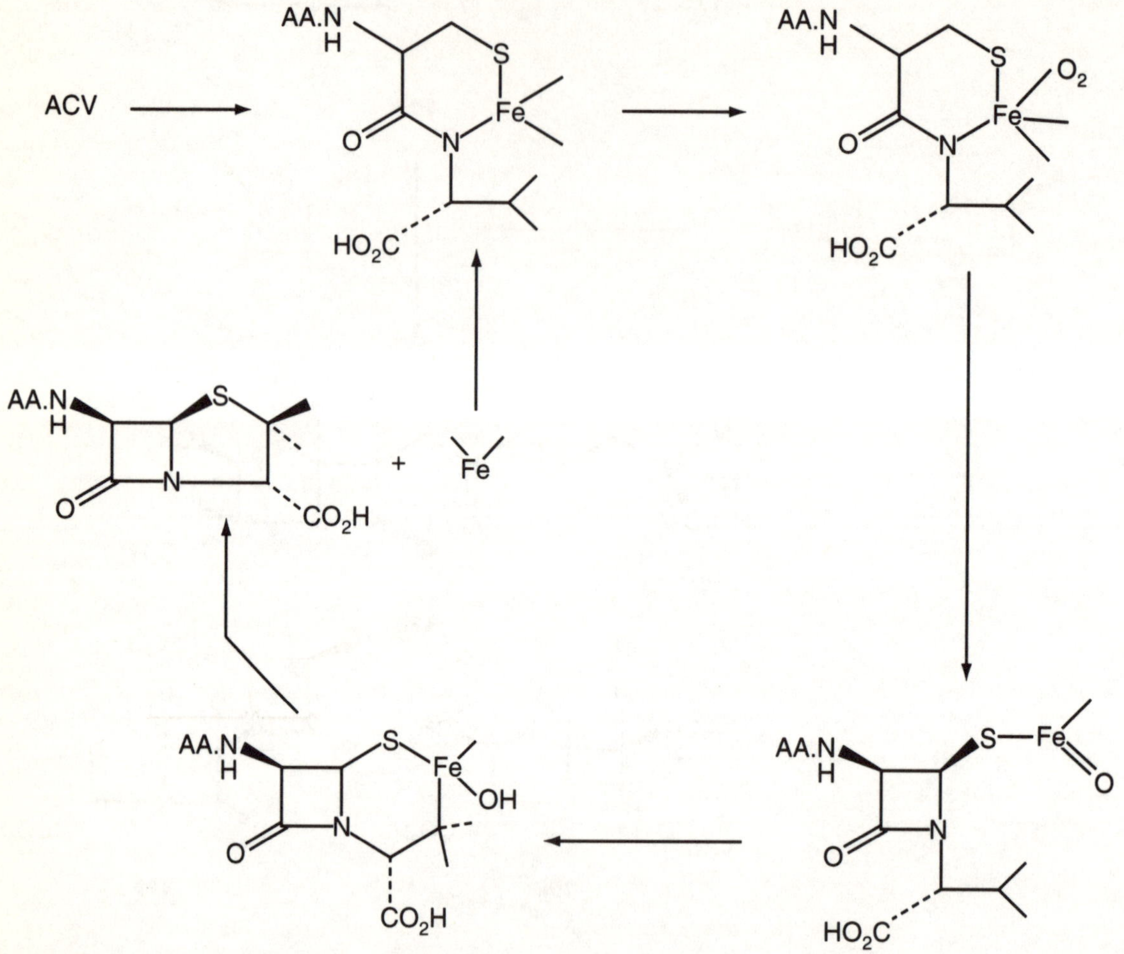

Scheme 52 The formation of the penicillin ring system

Analogue biosynthesis with variants on the tripeptide led to a range of novel β-lactam antibiotics. Thus, the use of *O*-methyl-D-allothreonine (**212**) in place of valine gave a 2α-methoxypenam (**213**) with a potentially useful biological activity.

The semisynthetic β-lactam antibiotics of the cephalosporin series are produced from 7-aminocephalosporanic acid (7-ACA), which is currently derived chemically from the natural cephalosporin C produced by the fungus *Cephalosporium acremonium*. Penicillin production in *Penicillium chrysogenum* involves replacement of the aminoadipyl side chain by phenoxyacetate or other acyl units that are susceptible to amidases and thus release the penicillin

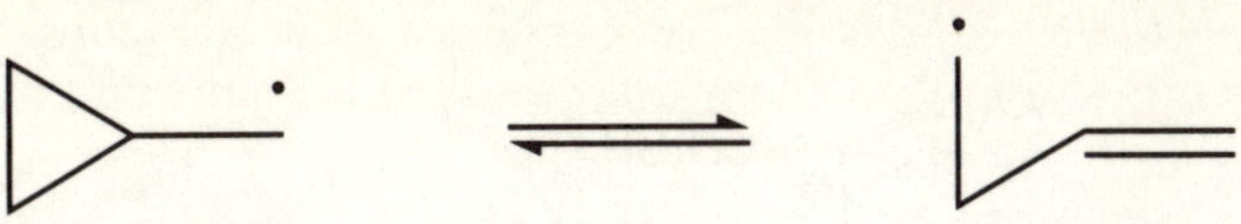

201 202

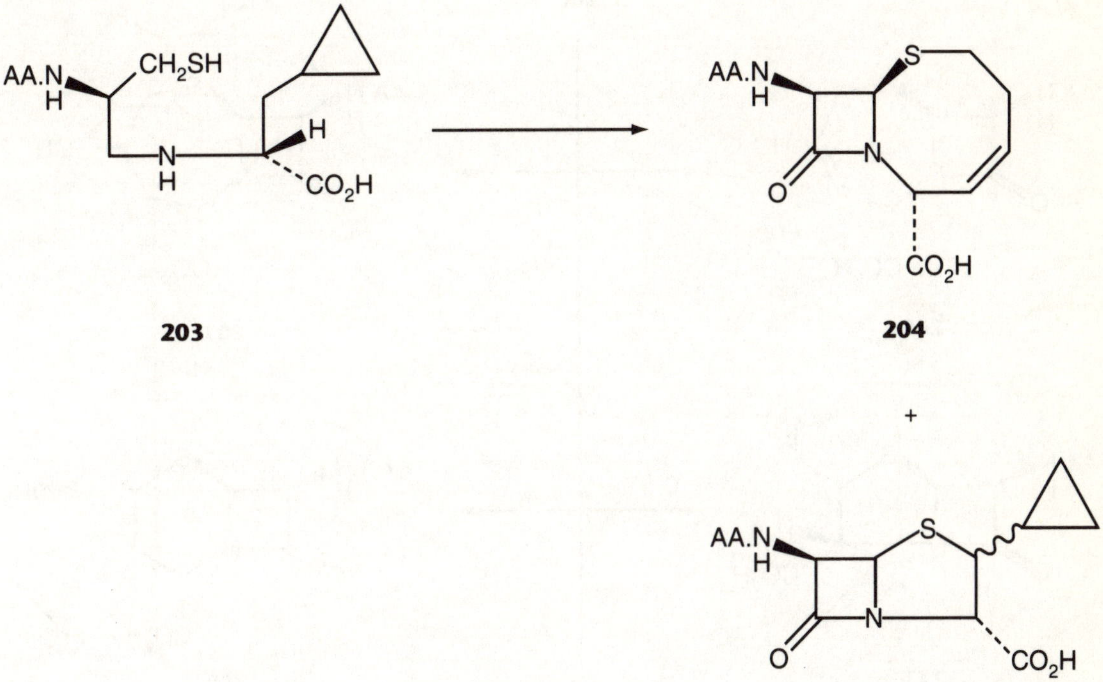

AA.N H CH$_2$SH N H H CO$_2$H

203

AA.N H S N O CO$_2$H

204

+

AA.N H S N O CO$_2$H

205

nucleus. However, a suitable cephalosporin C amidase capable of releasing 7-ACA enzymatically from cephalosporin C is not readily available. A combination of genetic engineering and analogue biosynthesis have overcome this problem.

Although the first steps in penicillin and cephalosporin biosynthesis are the same, the cephalosporins are produced by isomerization of the L-α-aminoadipate of isopenicillin N to the D-isomer and ring expansion of the penicillin N to the desacetoxycephalosporanic acid. Hydroxylation and acylation then afford cephalosporin C. The genes from *Cephalosporium acremonium* that code for the expandase, hydroxylase and acetyltransferase were expressed in a penicillin production strain of *Penicillium chrysogenum* and this mutant was grown in the presence of adipic acid to provide the side-chain. The organism then produced a novel metabolite, adipyl-7-ACA, which was susceptible to enzymatic cleavage by an amidase and provided the cephalosporin nucleus.

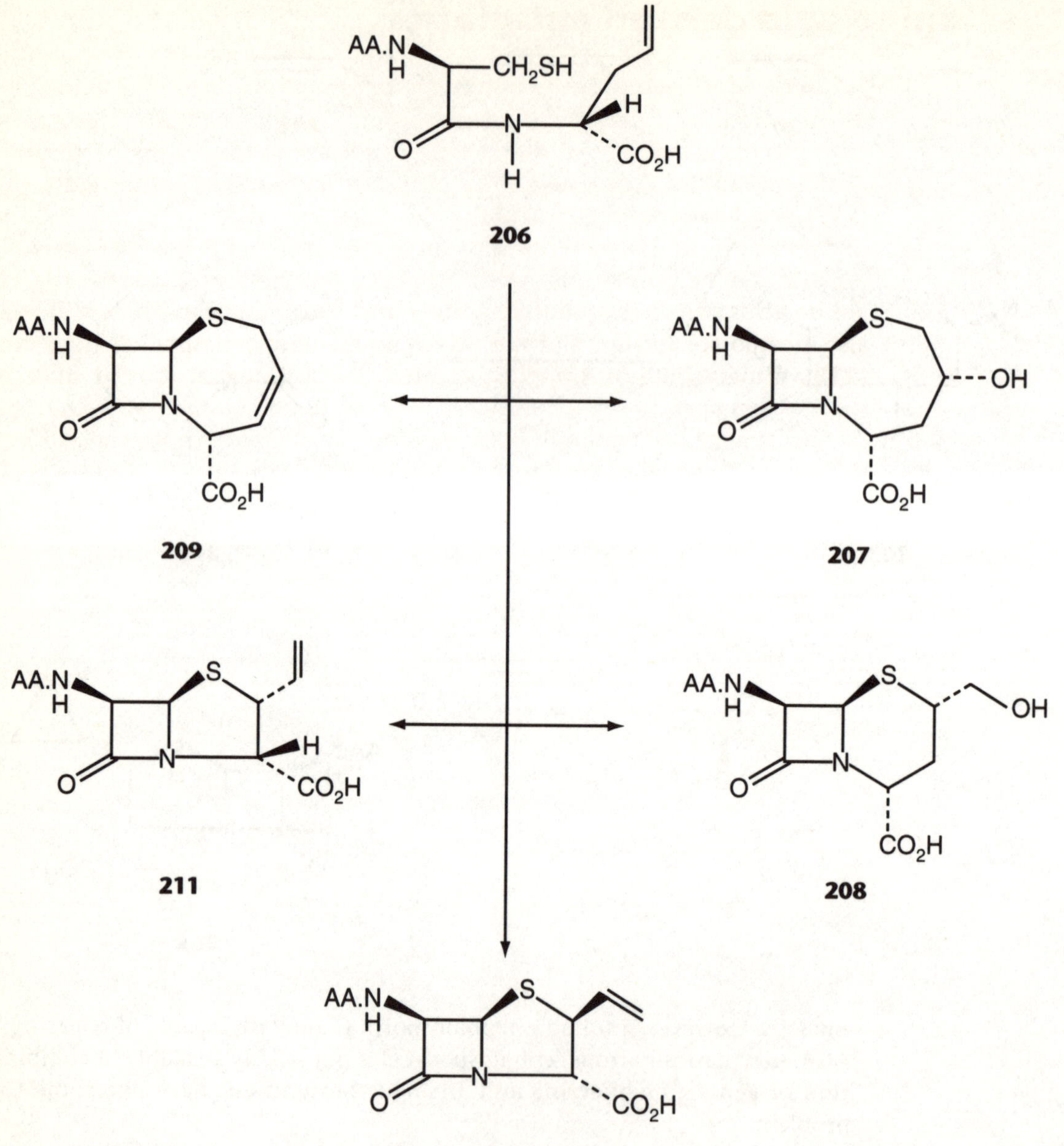

AA.N H CH$_2$SH H N H O CO$_2$H

206

AA.N H S N O CO$_2$H

209

AA.N H S OH N O CO$_2$H

207

AA.N H S H N O CO$_2$H

211

AA.N H S OH N O CO$_2$H

208

AA.N H S N O CO$_2$H

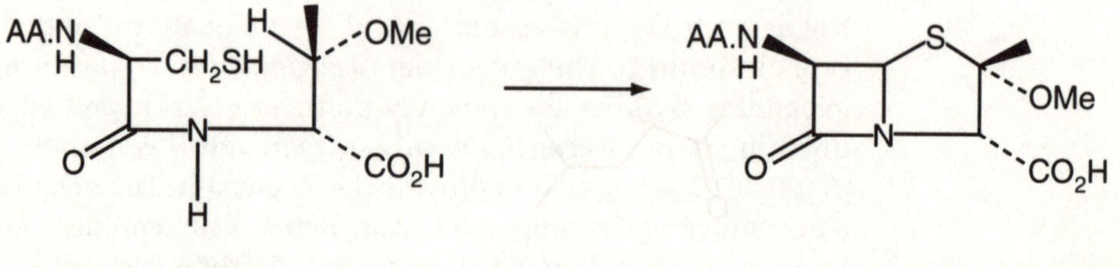

210

AA.N H CH$_2$SH H OMe N H O CO$_2$H

212

AA.N H S OMe N O CO$_2$H

213

Other amino-acid derived antibiotics

A number of antibiotics are derived from amino acids and their production can be influenced by the supply of exogeneous modified amino acids or peptides. Thus gliotoxin (**216**) is produced from the cyclic peptide cyclo-(L-seryl-L-phenylalanine) (**214**) by *Trichoderma viride*. Replacement of the serine by alanine in (**215**) afford 3a-deoxygliotoxin (**217**).

Cyclosporin A is a powerful immunosuppressant and is used clinically in organ and bone marrow transplants. It is a cyclic oligopeptide comprising 11 amino acids, which is produced by the fungus *Tolypocladium inflatum*. It will also incorporate 'foreign' amino acids to give modified cyclosporins. Asperlicin (**218**), a metabolite of *Aspergillus alliaceus*, is of biological interest since it specifically targets the cholecystokinin receptor. It is biosynthesized from tryptophan, leucine and two anthranilate residues. Analogues have been biosynthesized using modified tryptophan and leucine residues.

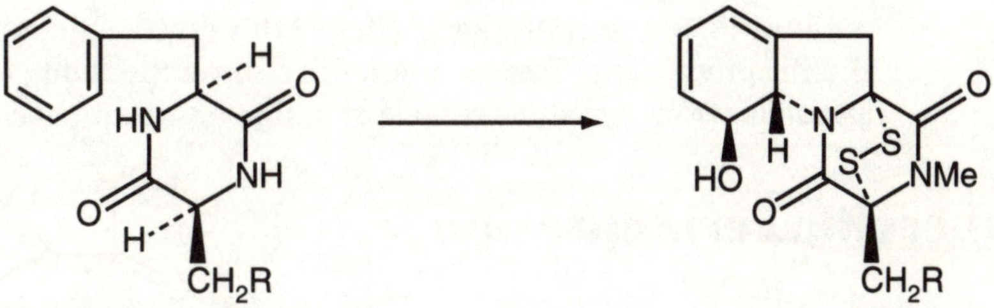

214 R = OH
215 R = H

216 R = OH
217 R = H

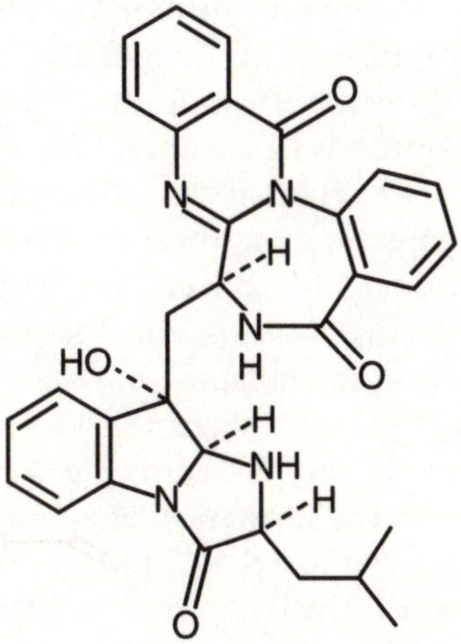

218

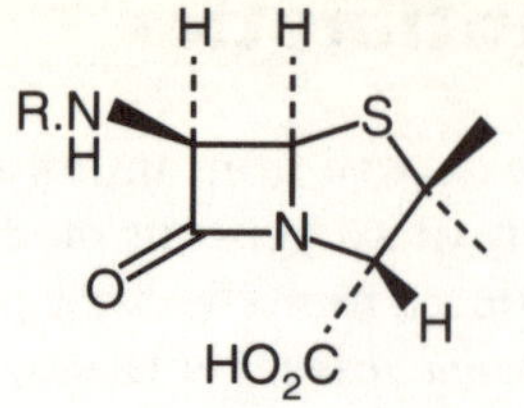

219 R = $CH_3CH_2CH{=}CHCH_2C(=O)$—

220 R = C_6H_5—$CH_2.C(=O)$—

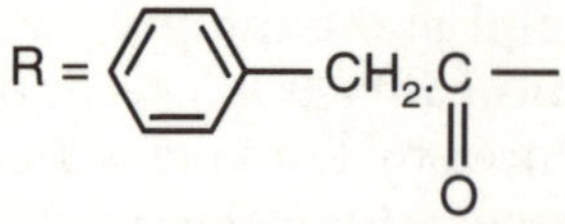

221 R = C_6H_5—$OCH_2.C(=O)$—

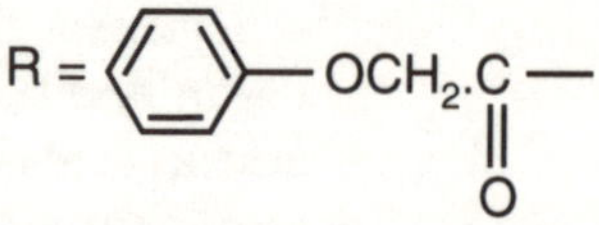

An early example of the use of micro-organisms to incorporate modified units intact into amino acid derived antibiotics comes from the penicillin fermentation mentioned earlier. The natural penicillins comprise a 6-APA unit to which a variable side chain is attached (e.g. (**219**) and (**220**)). By adding substituted precursors such as 2-p-fluorophenylethylamine or phenoxyacetic acid, modified penicillins (e.g. (**221**) were produced.

However, the successful application of this strategy to structure–activity studies and the production of novel antibiotics rests on the ability to separate the novel metabolite from its natural relative or to suppress the production of the latter.

The gibberellin fermentation

The fungus *Gibberella fujikuroi* produces gibberellic acid (**225**), the best known of the gibberellin plant hormones, as a major metabolite. It is a diterpenoid that is biosynthesized via ent-kaurene (**222**) and the acids (**223**) and (**224**), according to Scheme 53. The gibberellins, of which there are some 90 examples, are produced in very small amounts (μg/kg) by higher plants and mediate many aspects of plant growth and development. Although the broad outline of the biosynthetic pathway is identical in both the fungus and in higher plants, the order and detail of some events differ. For example, hydroxylation at C–13 is the last stage in gibberellic acid biosynthesis in the fungus but occurs at a much earlier stage in some plants. Because of the paucity of material from plants and their absence from the normal fungal system, there has been quite a bit of interest in devising syntheses of the rarer 13-hydroxy plant gibberellins in order to assess their biological activity and metabolism. The diterpenoid steviol, *ent*-13-hydroxykaurenoic acid (**226**) is readily available from the commercial sweetener, stevioside, and embodies both an early stage, the 19-carboxyl, and the last stage in gibberellic acid biosynthesis, the 13-hydroxyl group. Incubation of the acid with *Gibberella fujikuroi* led to the formation of a number of 13-hydroxygibberellins (e.g. (**227**), which are characteristic of higher plants. Other higher plant gibberellins have been produced by the microbiological transformation of variously hydroxylated ent-kaurenes and indeed this has been used in the proof of structure (e.g. (**228**), (**229**)).

Scheme 53

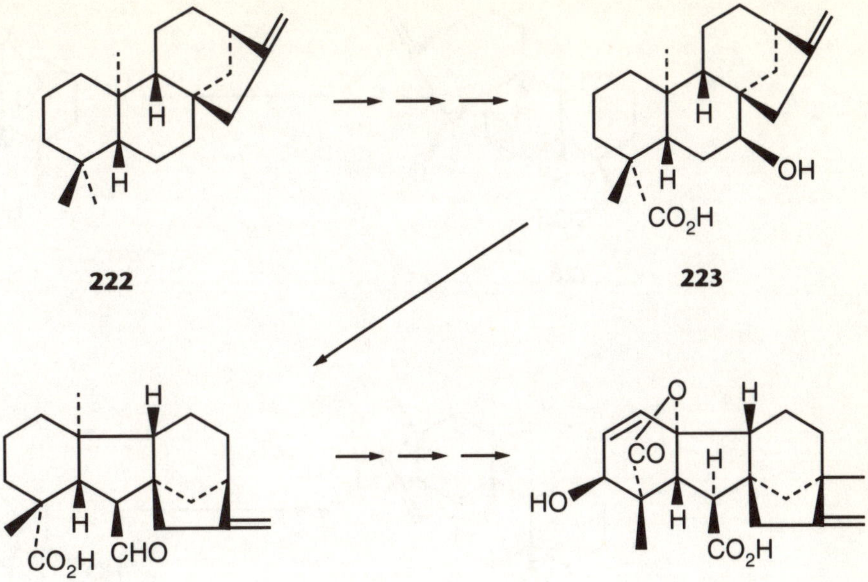

222 223

224 225

These biosynthetically-patterned transformations are not restricted to compounds with an ent-kaurene backbone. There are a number of closely related diterpenoid skeleta and representative examples of these have been used in biotransformations with *Gibberella fujikuroi*, e.g. (**230**) gave (**231**).

Gibberellic acid (**225**) is chemically quite labile, undergoing both rearrangement and decomposition in acid and alkali. Consequently, the preparation of labelled gibberellic acid and modified gibberellins bearing the reactive functionality can be quite difficult. Microbiological transformation has been used to overcome this problem. Thus, the alcohol (**232**), which is related to the known intermediate gibberellin A_{12} aldehyde (**224**), was readily prepared, labelled at C-17. Incubation with *G. fujikuroi* afforded labelled gibberellic acid. The alkylgibberellin (**233**) was more readily prepared and then converted to its gibberellic acid analogue by biotransformation.

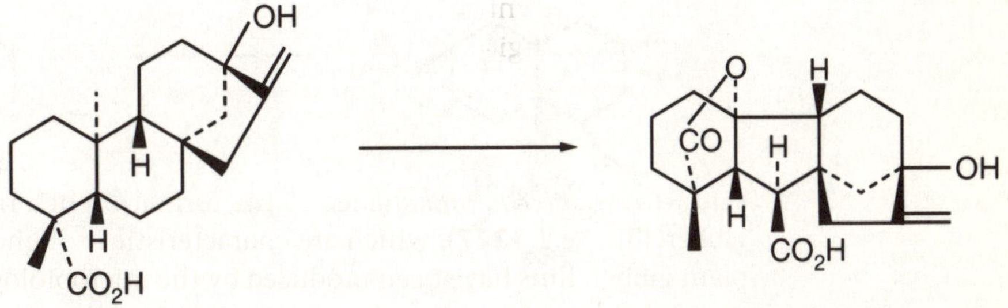

226 227

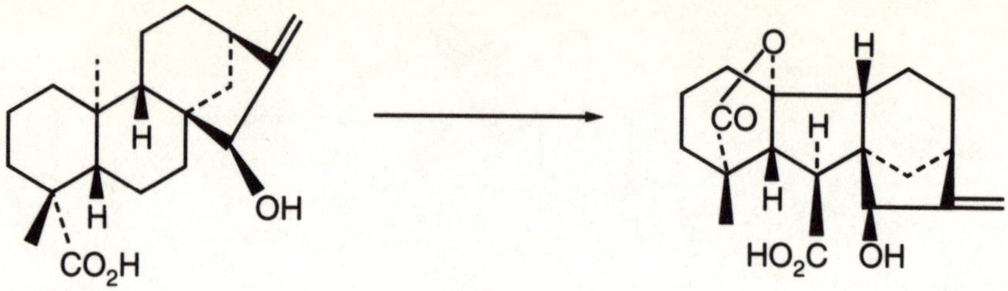

228 **229**

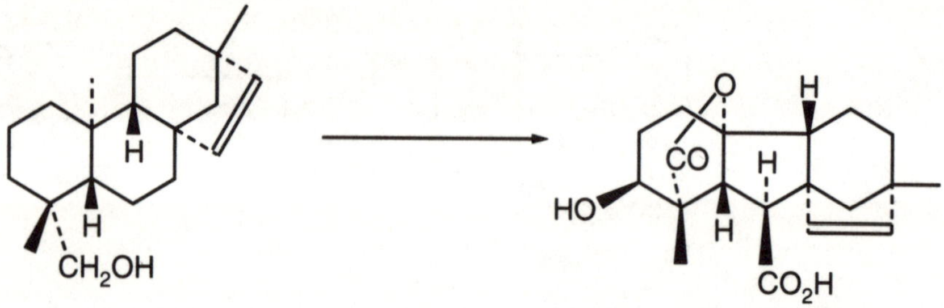

230 **231**

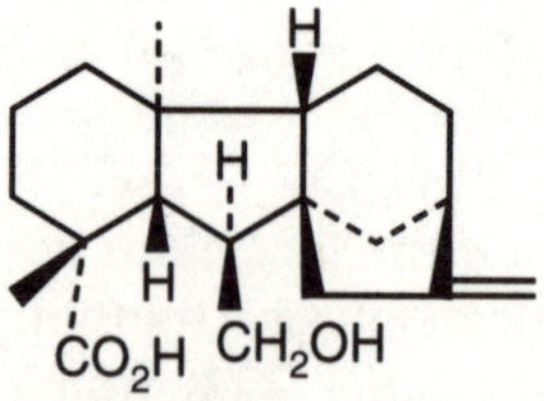

232 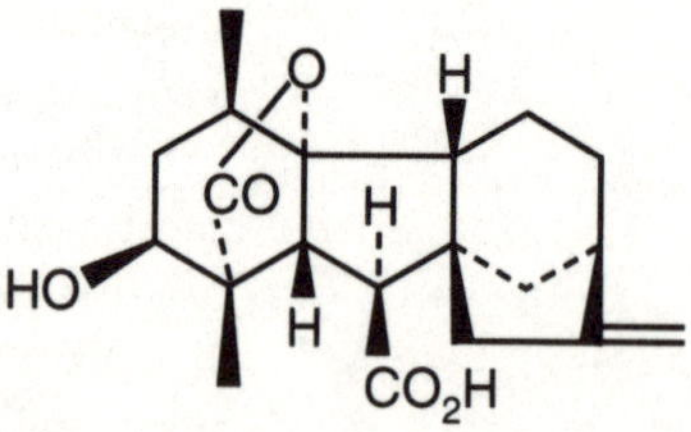 **233**

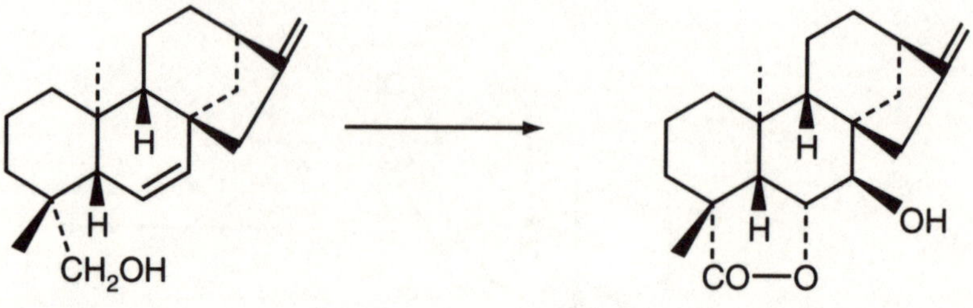

234 **235**

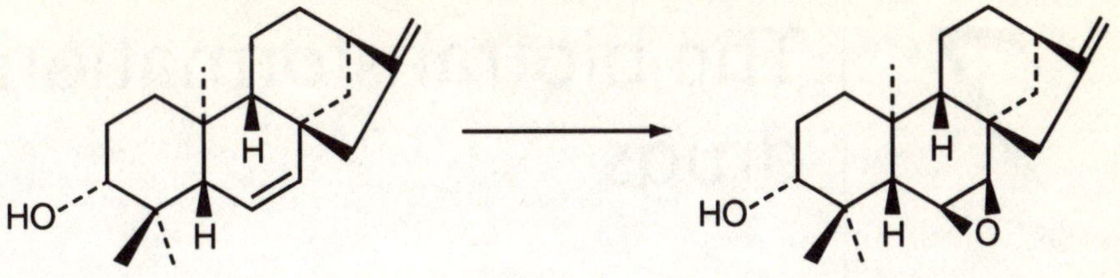

236 237

The results of these biotransformations revealed a number of interesting constraints. An additional hydroxyl group adjacent to some centres, that were normally metabolized, inhibited that particular step. Thus hydroxylation at C–19 of *ent*-kaurene is inhibited by the presence of an 'additional' 3α-hydroxyl group, whilst the ring-contraction step involving oxidation at C–6 is inhibited by a hydroxyl group at C–18. This was put to biosynthetic use. Epoxidation of the 6,7-ene in *ent*-kaur-6, 16-dienoic acid (**234**) was a probable step in the biosynthesis of the kaurenolide lactones (**235**) by the fungus. However, the epoxide was too reactive to trap in the presence of a 19-carboxylic acid. Incubation of *ent*-3α-hydroxy kaur-6, 16-diene (**236**), in which oxidation of the 19-methyl group was inhibited, gave the corresponding 6β, 7β-epoxide (**237**), providing some evidence for this step in the biosynthesis of the kaurenolide lactones.

7 The biotransformation of drugs

The study of the biotransformation of a drug, and in particular its metabolism in man, forms a major part of its pharmaceutical evaluation. The correlation of biotransformation with pharmacological activity and with toxicity and adverse side-effects are important in the development of modifications of the parent drug. The metabolism of a drug may produce its biological activity. Use is made of this in the concept of a pro-drug. A pro-drug is a derivative of a drug that may have enhanced stability, absorption or transport characteristics or masked side-effects and from which the parent active drug is released, preferably at the site of action, affording a prolonged and target-specific activity. On ingestion, a drug passes from the mouth to the stomach and the gastrointestinal tract, from whence it is adsorbed into the bloodstream and transferred via the liver and the lung eventually to the site of action. There are many barriers and sites of biotransformation on this route.

As a rough generalization, metabolism in the mouth, stomach and gastrointestinal tract, whether by the digestive enzyme system or by gut micro-flora, tends to be hydrolytic and reductive, particularly in the anaerobic situation of the bowel. In contrast, metabolism in the liver and the lung tends to be oxidative and involves conjugation, the attachment of polar water solubilizing residues. Such biotransformations, particularly in the liver, may lead to excretion of a drug before it has produced its biological effect. This is known as a 'first-pass effect' or 'first-pass loss'.

The metabolic changes are of two types. Phase I reactions include hydroxylation, oxidation, reduction and hydrolysis. Phase II reactions involve 'conjugation' – that is, the addition of a second group such as glucuronic acid or a sulphate moiety. We will consider examples of these changes in the following sections.

Phase I changes: aromatic hydroxylation

Many of the phase I changes have been encountered in earlier chapters and hence the discussion here will be limited to those aspects which affect specific drugs and the development of toxicity.

Many drugs contain an aromatic ring that undergoes hydroxylation by the mixed function oxidases found, for example, within the hepatic endoplasmic reticulum. These cytochrome P_{450}-dependent enzymes carry out many transformations. The oxidation of an aromatic ring involves the formation of an arene-oxide (**238**). This reactive species may undergo rearrangement with the migration of a hydrogen atom (the NIH shift) and the formation of a phenol (**240**) via its ketonic form. Alternatively, the epoxide may be cleaved with the addition of a nucleophile leading, for example, to the formation of a *trans*-diol (**239**). The nucleophile may be the thiol of glutathione (**242**) affording a conjugate (**241**). Alternatively, and more seriously, the nucleophile may be a basic residue on a protein or nucleic acid. In these situations the drug may become attached to a protein or nucleic acid fragment and produce permanent damage

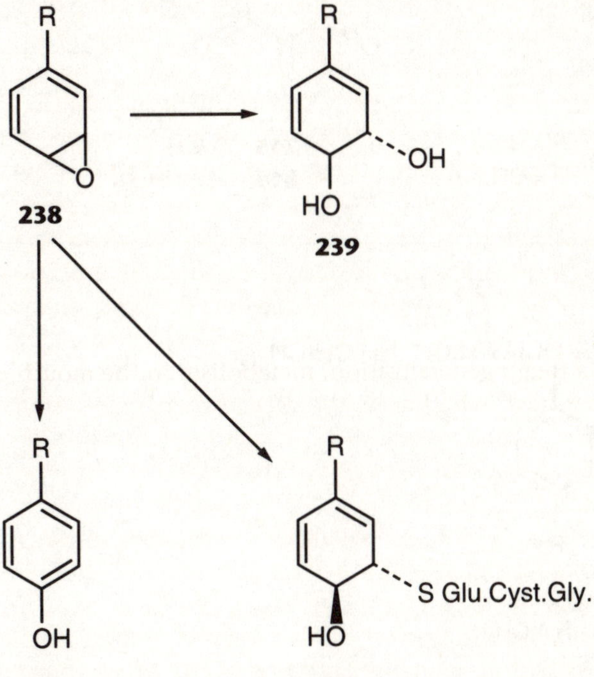

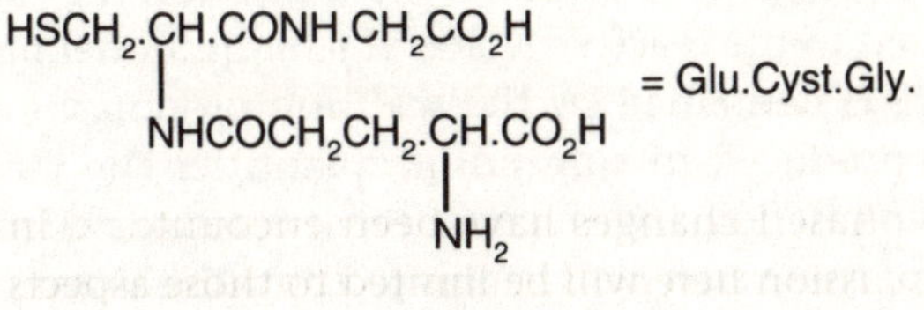

R

OH

HO

238

239

R

R

S Glu.Cyst.Gly.

OH

HO

240

241

HSCH$_2$.CH.CONH.CH$_2$CO$_2$H

NHCOCH$_2$CH$_2$.CH.CO$_2$H = Glu.Cyst.Gly.

NH$_2$

242

to the tissue, e.g. hepatotoxicity or DNA damage. The biotransformation of amphetamine (**243**) exemplifies these points. Characteristically, the hydroxyl group that is inserted (**244**) appears in the *para* position to the original directing substituent. Another example is the metabolism of the sedative phenobarbitone (**245**) to p-hydroxy phenobarbitone (**246**) and the β-blocker propranolol (**247**) to 4-hydroxy-propranolol (**248**).

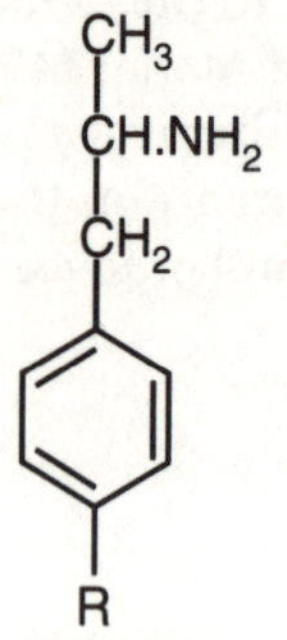

CH_3
$CH.NH_2$
CH_2
R

243 R = H
244 R = OH

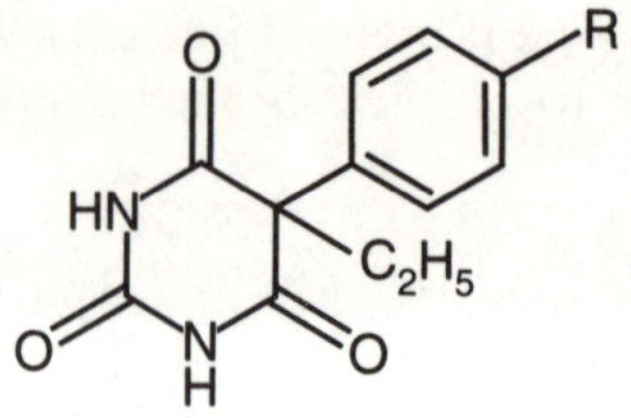

O R HN C_2H_5 O N H O

245 R = H
246 R = OH

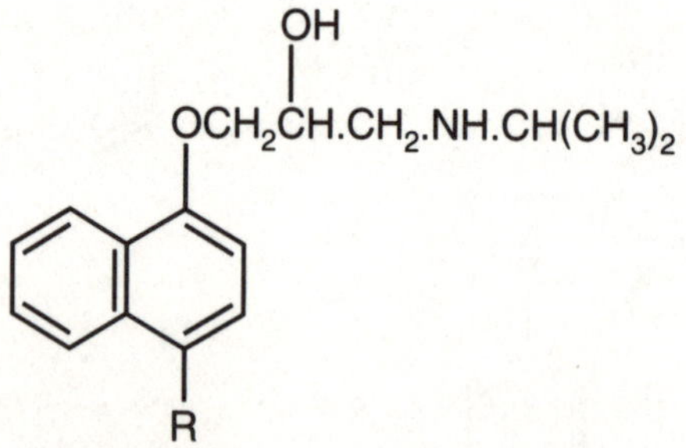

OH
$OCH_2CH.CH_2.NH.CH(CH_3)_2$
R

247 R = H
248 R = OH

The carcinogenicity that is associated with polycyclic aromatic hydrocarbons such as benzpyrene (**249**) arises from their biotransformation to arene-oxides and the subsequent cleavage of these to afford compounds such as (**250**). Some phenols may be further oxidized to the quinones, for example paracetamol (**251**) is oxidized to the reactive *N*-acetyl-*para*-quinoneimine (**252**). Although this metabolite is detoxified by the addition of glutathione (**242**), to afford (**253**), if it is produced in substantial quantities the reaction with cellular macromolecules can occur, with consequent cytotoxic effects.

The ability of these enzyme systems to epoxidize unsaturated systems is not restricted to the formation of arene-oxides. Thus, double bonds such as that in carbamezipine (**254**) are epoxidized to form (**255**).

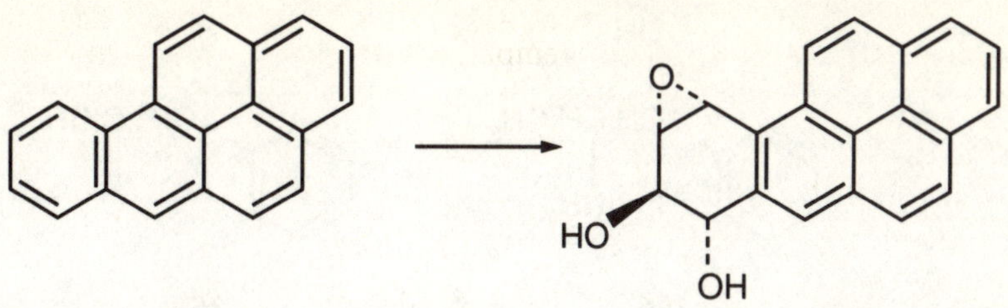

249 250

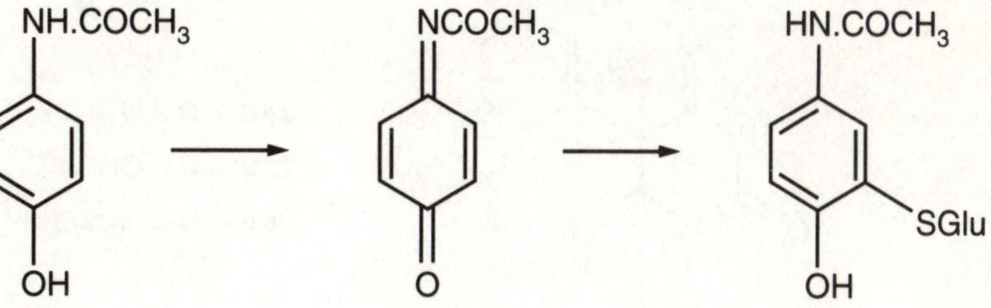

251 252 253

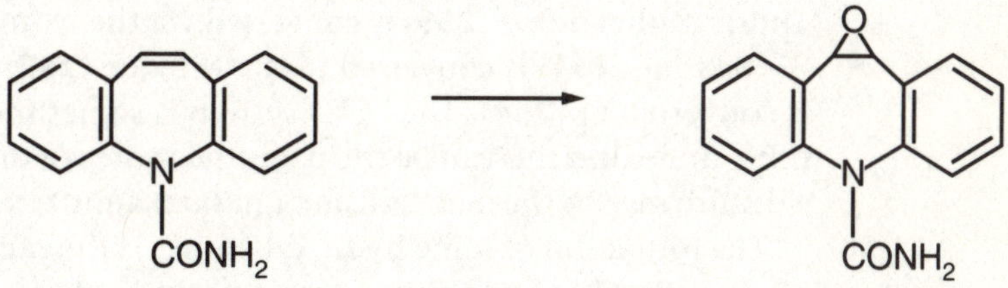

254 255

Hydroxylation at benzylic and allylic positions

Hydroxylation at benzylic positions is another common biotransformation. This may be seen as a position that is sensitive to radical attack. A typical example is the conversion of amphetamine (**243**) to the benzylic alcohol (**256**). There is an interesting structural similarity between this metabolite and nor-ephedrine. Another example of this type of metabolic hydroxylation is seen in the conversion of hexobarbital (**258**) via the allylic alcohol (**259**) to the ketone (**260**).

Oxidation of hetero-atoms

The oxidation of nitrogen and sulfur functions is another common biotransformation of drugs. The monoamine oxidase (MAO) system is responsible for a number of changes, including the oxidation of tertiary amines to their

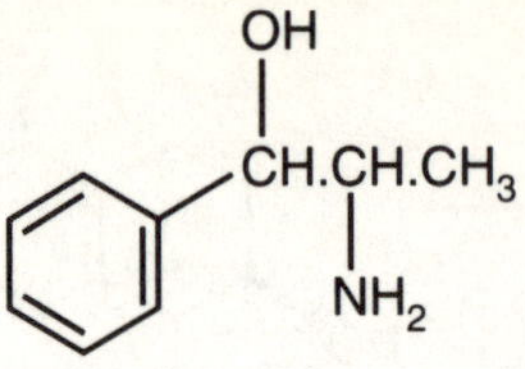

256

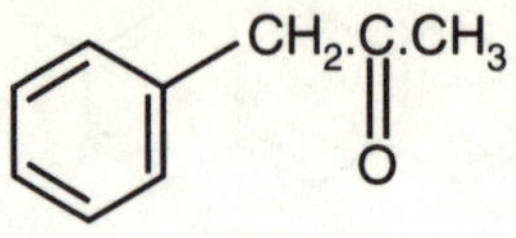

257

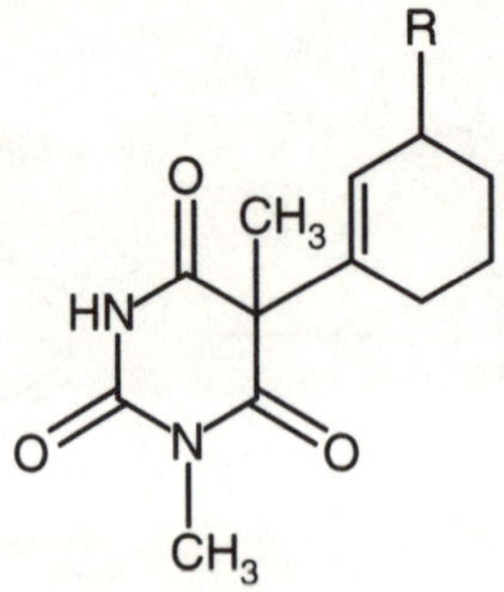

258 R = H

259 R = OH

260 R = =O

N-oxides and of primary and secondary amines to oximes and hydroxylamines. Thus, amphetamine (**253**) is converted via the oxime to the ketone (**257**).

Nicotine (**261**) is converted to the *N*-oxide (**262**) and chlorpromazine (**263**) is converted to (**264**). The MAO system is sometimes deficient or inhibited by other drugs and this can be the potential source of drug:drug interactions and of a disturbance in the metabolism of natural amines such as the biogenic amines.

The formation of some hydroxylamines is the cause of toxicity. For example, phenacetin (**266**) was at one time widely used as a painkiller, but it is metabolized in part to the *N*-hydroxy derivative (**267**), which is a toxic metabolite (Scheme 54).

Sulfur atoms are also oxidized to the sulfoxide and sulfone. For example, the phenothiazine chlorpromazine is oxidized to the sulfone (**265**) as well as being hydroxylated on the aromatic ring at C–7 and C–8. Indeed, these hydroxylation products may cross the blood–brain barrier and might also be involved in the psychotropic activity.

Dealkylation

The dealkylation of hetero-atoms is also a common metabolic process. This may occur oxidatively. Thus, phenacetin (**266**) is also a pro-drug for paracetamol (**251**). *N*-Dealkylation is also found in many situations. Imipramine (**268**) is an antidepressant that is metabolized to its *N*-desmethyl compound desipramine (**269**), which is also an antidepressant drug but with a slightly different spectrum of activity.

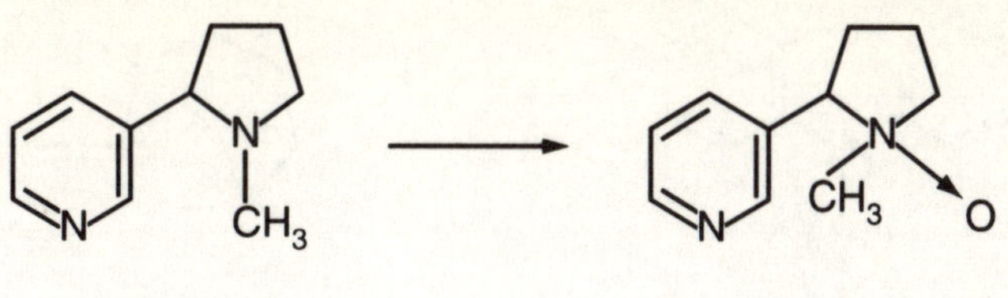

261 262

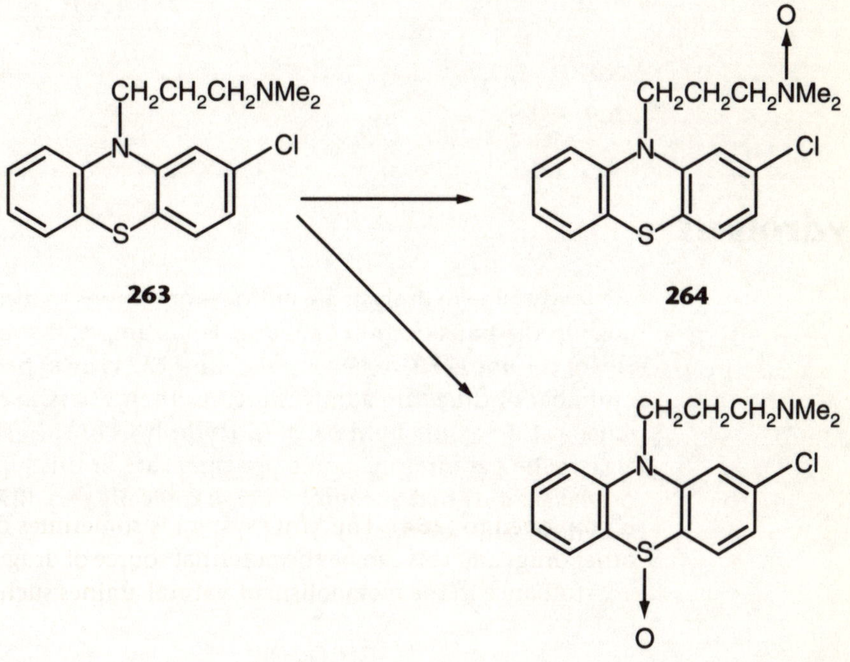

263 264

265

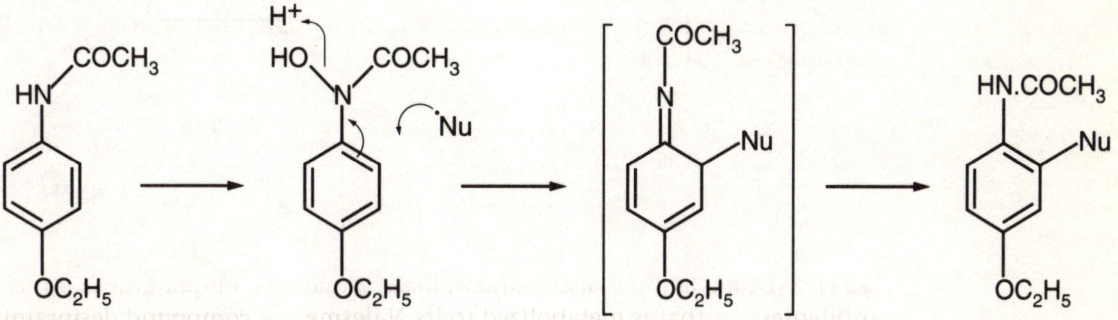

266 267

Scheme 54 The metabolism of phenacetin.

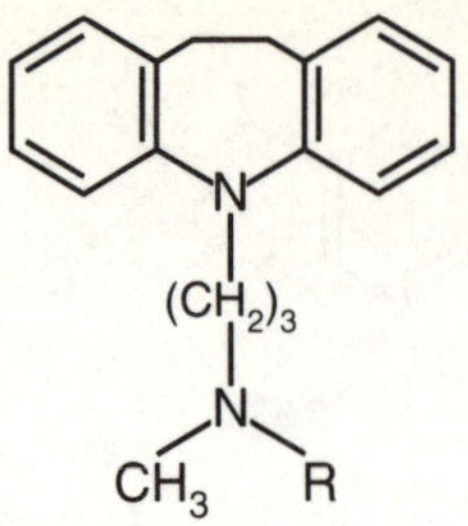

268 R = CH_3
269 R = H

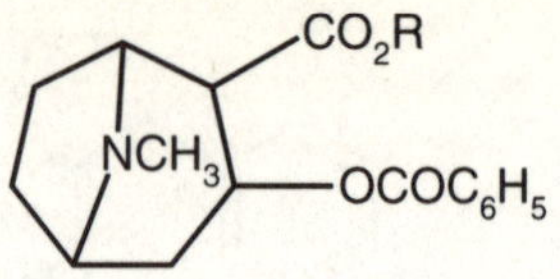

270 R = CH_3
271 R = H

Hydrolysis

Ester and amide hydrolysis are biotransformations that can occur at a number of stages in the translocation of a drug. For example, the hydrolysis of the methyl ester of cocaine (**270**) to benzoylecgonine (**271**) takes place in the blood plasma. A number of drugs are administered as their esters in order to facilitate their translocation across lipid barriers. Hydrolysis of the ester by an esterase then releases the parent drug. Some pro-drugs are of this type. Thus, several of the cortical steroids that are administered topically (e.g. (**272**)) are esters that may be hydrolysed.

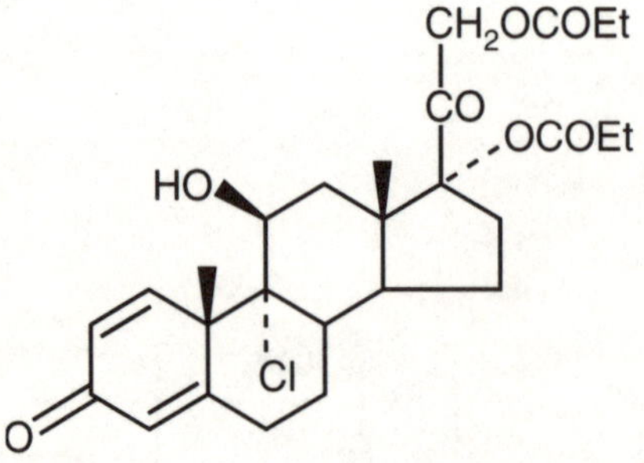

272

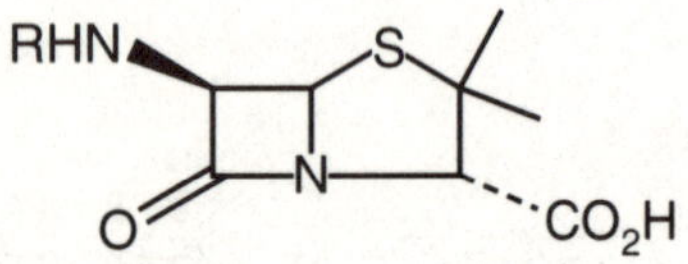

273

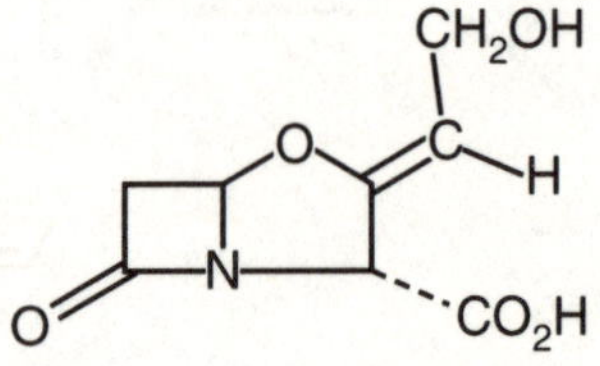

274

Many of the penicillin-resistant strains of bacteria have evolved efficient β-lactamases that cleave the β-lactam ring of the penicillin (**273**) and thus destroy its activity. A natural product, clavulanic acid (**274**), has been found that inhibits the β-lactamases and, in combination with various penicillins, allows them to express their activity.

Reduction

Although many of the changes that occur to a drug are oxidative, there are some significant reductive steps. The reduction of a nitro group to an amino group, particularly by the gut micro-flora, can significantly alter the uptake of a drug. Thus the nitro group in the antibiotic chloramphenicol (**275**) is reduced to an amine (**276**), which can form a salt in the acidic region of the stomach. Similarly, the tranquillizer nitrazepam (**277**) is converted to the amino-derivative (**278**).

Am important biotransformation led to the discovery of the sulfonamide drugs. The azo-dyestuff Prontosil Red (**279**) was tested in rabbits infected with various bacteria. It proved to be an effective anti-bacterial agent although it was relatively ineffective in simple *in vitro* assays. The activity was traced to its metabolite, sulphanilamide (**280**), and it is from this substance that the important family of sulfonamide antibiotics are derived.

Phase II changes: conjugation

The major conjugation reactions that lead to the deactivation and excretion of drugs involve glucuronide formation, glutathione conjugation, glycine and

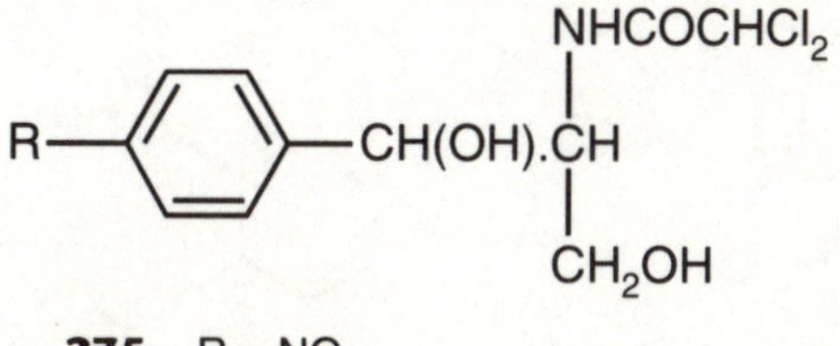

$NHCOCHCl_2$

R—C₆H₄—CH(OH).CH—CH_2OH

275 $R = NO_2$
276 $R = NH_2$

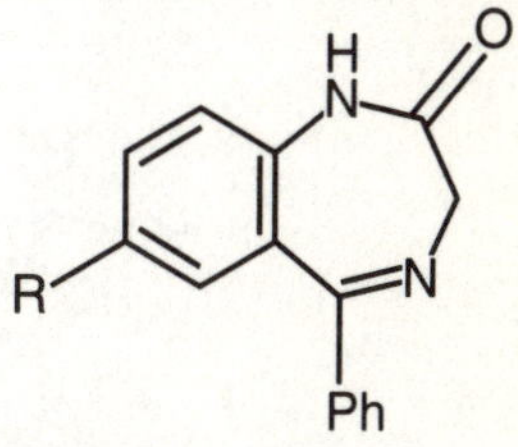

Ph

277 $R = NO_2$
278 $R = NH_2$

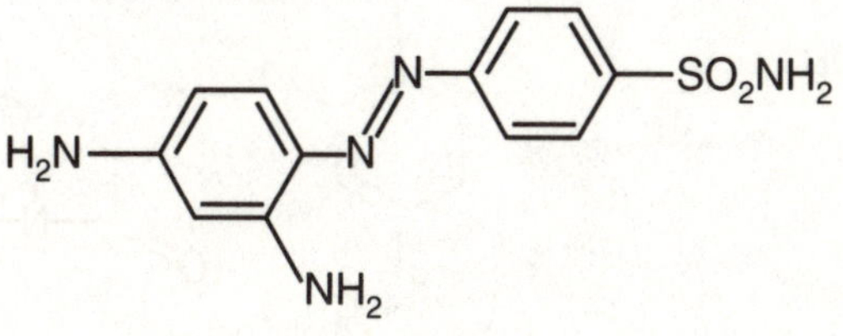

H_2N N N SO_2NH_2 NH_2

279

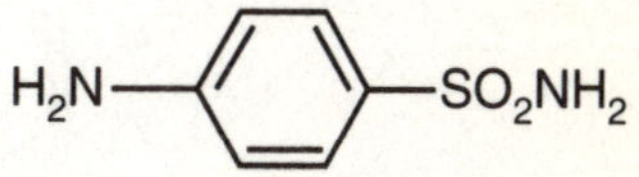

H_2N—C₆H₄—SO_2NH_2

280

sulfate conjugation. Glucuronic acid conjugation, particularly of hydroxyl groups, is a very common biotransformation. It involves the transfer of glucuronic acid (**281**) from uridine diphosphate glucuronic acid (UDPGA) to the substrate mediated by the enzyme UDP glucuronyl transferase. Many alcohols, phenols and acids are excreted as their glucuronides, e.g. aspirin (**282**) as (**283**).

Carboxylic acids are also conjugated to the amino acid glycine. For example, the tuberculostatic agent, *p*-aminosalicylic acid (**284**) is converted to the glycine conjugate (**285**) and to the glucuronide (**286**). These changes can be seen to increase the polarity and water solubility of the metabolites and thus diminish their absorption.

The amino acid glycine forms part of the tripeptide glutathione (**242**). The nucleophilic thiol grouping of the cysteine moiety can react with a variety of electron-deficient centres. It regularly becomes attached to aromatic rings through the opening of an arene-oxide. Glutathione conjugation often acts as a protective mechanism against the harmful effects of the arene-oxide metabolic pathway.

Sulfate conjugation involves the esterification of a hydroxyl group with the sulfate moiety to give a highly polar ionized product with a marked water solubility. The sulfate is derived from the 3′-phospho-adenosine-5′-phosphosulfate in a biotransformation catalysed by sulfo-transferase. Typical sulfates are formed by estrone (**287**) and by simpler molecules such as (**288**). Although the majority of sulfates are rapidly excreted, the sulfate group is a good leaving group, generating electron-deficient species that may react with nucleophilic centres in cellular macromolecules. Thus, some carcinogenic aromatic amines such as β-naphthylamine are converted to the *N*-hydroxy derivatives and it is the sulfates of these that react with DNA. The flavouring safrole (**289**) is

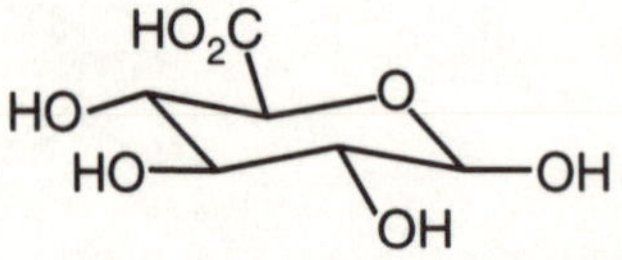

HO_2C HO HO OH OH

281

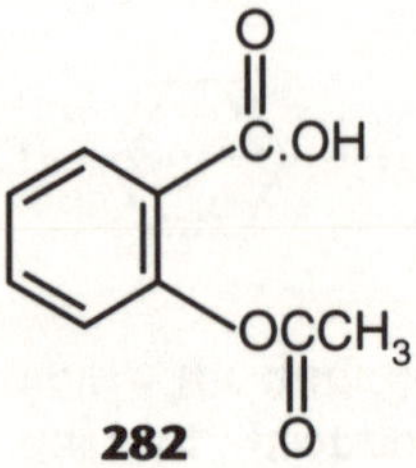

C.OH $OCCH_3$ O

282

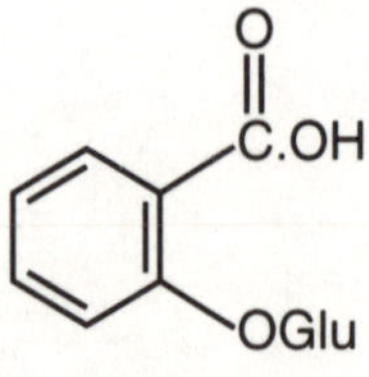

C.OH OGlu

283

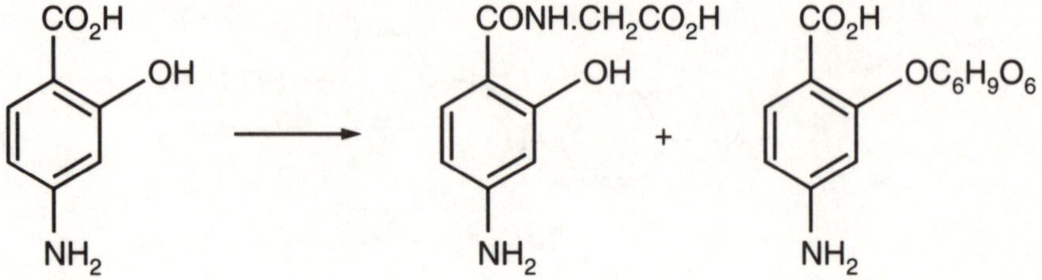

CO_2H OH NH_2

284

$CONH.CH_2CO_2H$ OH NH_2

285

+

CO_2H $OC_6H_9O_6$ NH_2

286

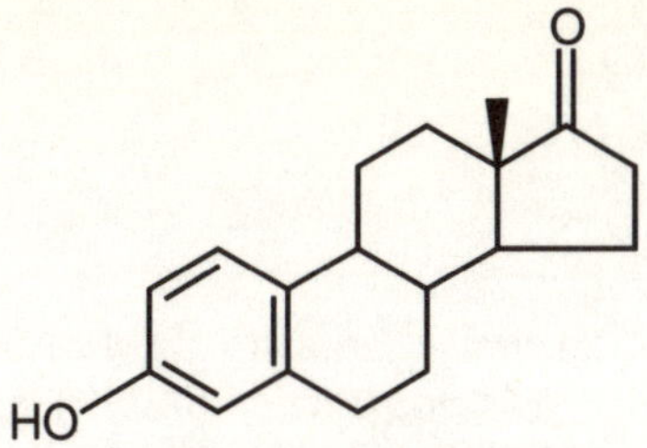

hydroxylated at the benzylic C–1 position. Sulfation of this hydroxyl group generates a very labile *O*-sulfate, which can react with cellular nucleophiles.

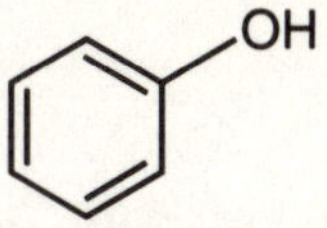

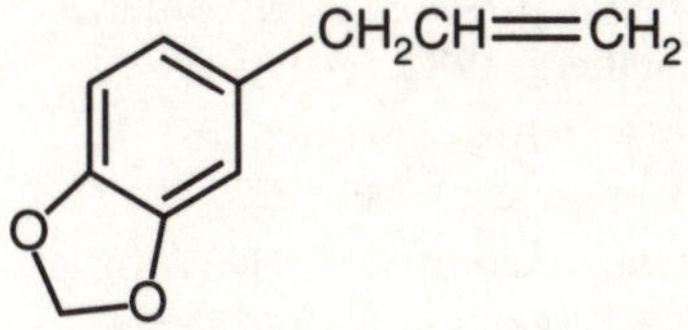

The majority of the studies on drug metabolism have been carried out with small animals, monkeys and man. There are considerable pressures to reduce the extent of animal testing and these have to be balanced against a need to establish a drug's metabolic fate and the biological activity of its metabolites in order to assess the safety of the drug. Whilst the use of biotransformation by micro-organisms as models for drug metabolism cannot completely overcome the need for mammalian studies, it can nevertheless provide an insight into the possible metabolic profile of a drug and yield quantities of a metabolite for characterization, identification and testing purposes. A number of micro-organisms have been examined in this context. For example, the metabolism of amphetamines by *Cunninghamella echinulata* has been found to parallel mammalian metabolism, whilst mammalian metabolites of various hormonal steroids are produced by the microbiological hydroxylation of progesterone and testosterone.

It can thus be seen that the biotransformation of drugs represents an important facet of the study of their physiological role.

Further reading

Books

Biotransformations in Preparative Organic Chemistry, H. G. Davies, R. H. Green, D. R. Kelly and S. M. Roberts, Academic Press, London, 1989.

Preparative Biotransformations, ed. S. M. Roberts, John Wiley, Chichester, 1992.

Biotransformations in Organic Chemistry, K. Faber, Springer-Verlag, Heidelberg, 1992.

Organic Synthesis with Oxidative Enzymes, H. L. Holland, VCH Publishers, New York, 1992.

Enzymes in Synthetic Organic Chemistry, C. H. Wong and G. M. Whitesides, Pergamon, Oxford, 1994.

Biotransformations, vol. 1–6, D. R. Hawkins, Royal Society of Chemistry, London, 1988–1994.

Microbiological Transformation of Steroids, W. Charney and H. L. Herzog, Academic Press, New York, 1968.

Biotransformation of Non-steroidal Cyclic Compounds, K. Kieslich, Georg Thieme, Berlin, 1976.

Reviews and papers

There are a large number of reviews devoted to various aspects of biotransformation. The following list is not intended to be exhaustive but may be useful as a basis for further reading.

General reviews

Enzymes as catalysts in synthetic organic chemistry, G. M. Whitesides and C. H. Wong, *Angew. Chem., Int. Ed.*, 1985, **24**, 617.

Enzymes in Organic Synthesis, J. B. Jones, *Tetrahedron*, 1986, **42**, 3351.

Recent advances in the use of enzyme-catalysed reactions in organic research, S. Butt and S. M. Roberts, *Nat. Prod. Rep.*, 1986, **3**, 489.

Opportunities for using enzymes in organic synthesis, S. Butt and S. M. Roberts, *Chem. Brit.*, 1987, 127.

Re-routing nature, A. I. Scott, *Chem. Brit.*, 1993, 687.

Catalytic Antibodies: The generation of novel biocatalysts, G. M. Blackburn and P. Wentworth, *Chem. Ind.*, 1994, 338.

Enzymes in organic synthesis, A. Akiyama, M. Bednarski, M. J. Kim, E. S. Simon, H. Waldmann and G. M. Whitesides, *Chem. Brit.*, 1987, 645.

Enzyme Chemistry, N. J. Turner, *Ann. Rep. Prog. Chem.*, (Section B), 1988, **85**, 307; 1989, **86**, 307; 1990, **87**, 333; A.G. Sutherland, 1991, **88**, 263; 1992, **89**, 281.

Recent advances in the use of enzyme-catalysed reactions in organic synthesis, N. J. Turner, *Nat. Prod. Rep.*, 1989, **6**, 625; 1994, **11**, 1.

Microbial and enzymatic processes for the production of biologically and chemically useful compounds, H. Yamada and S. Shimizu, *Angew. Chem., Int. Ed.*, 1988, **27**, 622.

Industrial synthesis of optically active compounds, R. Sheldon, *Chem. Ind.*, 1990, 212.

Industrial developments in biocatalysis, V. H. M. Elferink, D. Breitgoff, M. Kloosterman, J. Kamphuis, W. J. J. van den Tweel and E. M. Meijer, *Recl. Trav. Chim. Pays-Bas*, 1991, **110**, 63.

Chapter 2 Hydrolytic reactions

Basic principles of protease-catalysed peptide bond formation, H. D. Jakubke, P. Kuhl and A. Konnecke, *Angew. Chem., Int. Ed.*, 1985, **24**, 85.

Active-site model for interpreting and predicting the specificity of pig liver esterase, E. J. Toone, M. J. Werth and J. B. Jones, *J. Am. Chem. Soc.*, 1990, **112**, 4946.

A concluding specification of the dimensions of the active site model of pig liver esterase, L. Provencher and J. B. Jones, *J. Org. Chem.*, 1994, **59**, 2729.

Pseudomonas fluorescens lipase in asymmetric synthesis, Z. F. Xie, *Tetrahedron Asymmetry*, 1991, **2**, 733.

Asymmetric transformations catalysed by enzymes in organic solvents, A. M. Klibanov, *Acc. Chem. Res.*, 1990, **23**, 114.

General aspects and optimization of enantioselective biocatalysis in organic solvents: The use of lipases, C. S. Chen and C. J. Sih, *Angew. Chem., Int. Ed.*, 1989, **28**, 695.

A rule to predict which enantiomer of a secondary alcohol reacts faster in reactions catalysed by cholesterol esterase, lipase from *Pseudomonas cepacia* and lipase from *Candida rugosa*, R. J. Kazlauskas, A. N. E. Weissfloch, A. T. Rappaport and L. A. Cuccia, *J. Org. Chem.*, 1991, **56**, 2656.

Microbial biotransformation of nitriles, L. A. Thompson, C. J. Knowles, E. A. Linton and J. M. Wyatt, *Chem. Brit.*, 1988, 900.

Application of nitrile converting enzymes for the production of useful compounds, T. Nagasawa and H. Yamada, *Pure Appl. Chem.*, 1990, **62**, 1441.

Chapter 3 Formation of C–C bonds

Use of enzymatic aldol reactions in synthesis, M. D. Bednarski, in *Comprehensive Organic Synthesis*, ed. B. M. Trost, Pergamon Press, Oxford, 1991, vol. 2, p. 455.

Enzyme catalysis in synthetic carbohydrate chemistry, D. G. Drueckhammer, W. J. Hennen, R. L. Pederson, C. F. Barbas, C. M. Gautheron, T. Krach and C. H. Wong, *Synthesis*, 1991, 499.

A combined chemical-enzymatic synthesis of 3-deoxy-D-arabinoheptulosonic acid 7-phosphate, N. J. Turner and G. M. Whiteside, *J. Am. Chem. Soc.*, 1989, **111**, 624.

Rabbit muscle aldolase as a catalyst in organic synthesis, M. D. Bednarski, E. S. Simon, N. Bischofberger, W. D. Fessner, M. J. Kim, W. Lees, T. Saito, H. Waldmann and G. M. Whitesides, *J. Am. Chem. Soc.*, 1989, **111**, 627.

Use of a recombinant bacterial fructose 1:6-diphosphate aldolase in aldol reactions; preparative syntheses of 1-deoxynojirimycin, 1-deoxymannojirimycin and 1, 4-dideoxy-1, 4-imino-D-arabinitol and fagamine, C. H. van der Osten, A. J. Sinskey, C. F. Barbas, R. L. Pederson, Y. F. Wang and C. H. Wong, *J. Am. Chem. Soc.*, 1989, **111**, 3924.

Fructose-1, 6-diphosphate aldolase catalysed stereoselective C–C bond formation, K. K. C. Liu, R. L. Pederson and C. H. Wong, *J. Chem. Soc. Perkin Trans. 1*, 1991, 2669.

A general approach to the synthesis of deoxy-azasugars, T. Kajimoto, L. Chen, K. K. C. Liu and C. H. Wong, *J. Am. Chem. Soc.*, 1991, **113**, 6678.

Rational design of azasugars via biocatalysis, T. Hudlicky, J. Rouden and H. Luna, *J. Org. Chem.*, 1993, **58**, 985.

Enzyme-catalysed synthesis of (*R*)-ketone cyanohydrins and their hydrolysis to (*R*)-ketone cyanohydrins and their hydrolysis to (*R*)-α-hydroxy-α-methylcarboxylic acids, F. Effenberger, B. Horsch, F. Weingart, T. Ziegler and S. Kuhner, *Tetrahedron Lett.*, 1991, **32**, 2605.

Preparation of chiral cyanohydrins by an oxynitrilase-mediated transcyanation, V. I. Ognyanov, V. K. Datcheva and K. S. Kyler, *J. Am. Chem. Soc.*, 1991, **113**, 6992.

Chapter 4 Redox reactions

Oxidation by microbial methods, S. M. Brown in *Comprehensive Organic Synthesis*, ed. B. M. Trost, Pergamon Press, Oxford, 1991, vol. 7, p. 53.

Reduction of C=X to CHXH using enzymes and micro-organisms, J. B. Jones, in *Comprehensive Organic Synthesis*, ed. B. M. Trost, Pergamon Press, Oxford, 1991, vol. 8, p. 183.

Baker's yeast as a reagent in organic synthesis, S. Servi, *Synthesis*, 1990, 1. Baker's yeast mediated transformations in organic chemistry, R. Czuk and B. I. Glanzer, *Chem. Rev.*, 1991, **91**, 49.

A new cubic-space reaction model for predicting the specificity of horse liver alcohol dehydrogenase catalysed oxido-reductions, J. B. Jones and I. G. Jakovac, *Can. J. Chem.*, 1982, **60**, 19.

Nicotinamide co-enzyme regeneration, J.B. Jones and K. E. Taylor, *Can. J. Chem.*, 1976, **54**, 2969.

Regeneration of deuteriated nicotinamide cofactors for use in large scale enzymatic synthesis of deuteriated substances, C. H. Wong and G. M. Whitesides, *J. Am. Chem. Soc.*, 1983, **105**, 5012.

Oxidative coupling of phenolic compounds, A. I. Scott, *Quart. Rev. Chem. Soc.*, 1965, **19**, 1.

Chapter 5 Microbiological hydroxylations

The microbiological degradation of aromatic compounds, D. W. Ribbons, *Ann. Rep. Prog. Chem.*, (Royal Society of Chemistry), 1965, **62**, 445.

New examples of microbial transformations in pharmaceutical chemistry, K. Kieslich, *Bull. Soc. Chim. (France)*, 1980, **11**, 9.

The microbiological hydroxylation of steroids and related compounds, E. R. H. Jones, *Pure Appl. Chem.*, 1973, **33**, 39.

The mechanism of the microbial hydroxylation of steroids, H. L. Holland, *Chem. Soc. Rev.*, 1983, 371.

Evidence for a carbon radical intermediate in the biohydroxylations achieved by the fungus, *Beauveria sulfurescens*, J. D. Fourneron, A. Archelas and R. Furstoss, *J. Org. Chem.*, 1989, **54**, 2478.

A unified mechanistic view of oxidative reactions catalysed by P_{450} and related Fe-containing enzymes, M. Akhtar and J. N. Wright, *Nat. Prod. Rep.*, 1991, 527.

Enzymatic hydroxylation of arenes and symmetry considerations in efficient synthetic design of oxygenated natural products, T. Hudlicky, R. Fan, H. Luna, H. Olivo and J. Price, *Pure Appl. Chem.*, 1992, **64**, 1109.

Bioconversion of sesquiterpenes, V. Lamare and R. Furstoss, *Tetrahedron*, 1990, **46**, 4109.

Microbiological hydroxylation of 5α-H-steroids, A. M. Turuta, N. E. Voishvillo and A. V. Kamernitskii, *Russ. Chem. Rev.*, 1992, **61**, 1033.

Current trends in microbial steroid biotransformation, S. B. Mahato and I. Majumdar, *Phytochemistry*, 1993, **34**, 883.

Side chain hydroxylation of aromatic compounds by fungi, exploring the benzylic hydroxylase of *Mortierella isabellina*, H. L. Holland, M. Kindermann, S. Kumaresan and T. Stefanac, *Tetrahedron Asymmetry*, 1993, **4**, 1353.

Chapter 6 Biosynthetically-directed biotransformations

The biosynthesis of penicillins and cephalosporins, J. E. Baldwin and E. P. Abraham, *Nat. Prod. Rep.*, 1988, **5**, 129.

New penicillins from isopenicillin N synthase, J. E. Baldwin, M. Bradley, S. D. Abbott and R. M. Adlington, *Tetrahedron*, 1991, **47**, 5309.

The folding of squalene, an old problem has new results, R. Bohlmann, *Angew. Chem., Int. Ed.*, 1992, **31**, 582.

Enzymatic cyclization of squalene and oxidosqualene to sterols and triterpenes, I. Abe, M. Rohmer and G. D. Prestwich, *Chem. Rev.*, 1993, **93**, 2189.

Plant cell culture combined with chemistry, J. P. Kutney, *Acc. Chem. Res.*, 1993, **26**, 559.

The microbiological transformation of diterpenoids, J. R. Hanson, *Nat. Prod. Rep.*, 1992, **9**, 139.

Biological variation of microbial metabolites by precursor directed biosynthesis, R. Thiericke and J. Rohr, *Nat. Prod. Rep.*, 1993, **10**, 265.

Chapter 7 The biotransformation of drugs

Impact of drug metabolism on medicinal research, D. V. Parke, *Chem. Ind.*, 1976, 380; 1976, 430.

Metabolites are important too, T. R. Marten, *Chem. Brit.*, 1985, 745.

Metabolic pathways, J. Caldwell and S. C. Mitchell in *Comprehensive Medicinal Chemistry*, ed. C. Hansch, P. G. Sammes and J. B. Taylor, Pergamon Press, Oxford, 1990, vol. 5, chapter 23.5.

Index